REENGINEERING THE SURVEY OF INCOME AND PROGRAM PARTICIPATION

Panel on the Census Bureau's Reengineered Survey of Income and Program Participation

Constance F. Citro and John Karl Scholz, *Editors*

Committee on National Statistics

Division of Behavioral and Social Sciences and Education

NATIONAL RESEARCH COUNCIL
OF THE NATIONAL ACADEMIES

THE NATIONAL ACADEMIES PRESS
Washington, D.C.
www.nap.edu

THE NATIONAL ACADEMIES PRESS 500 Fifth Street, N.W. Washington, DC 20001

NOTICE: The project that is the subject of this report was approved by the Governing Board of the National Research Council, whose members are drawn from the councils of the National Academy of Sciences, the National Academy of Engineering, and the Institute of Medicine. The members of the committee responsible for the report were chosen for their special competences and with regard for appropriate balance.

This study was supported by contract number YA132306CN0025 between the National Academy of Sciences and the U.S. Census Bureau. Support for the work of the Committee on National Statistics is provided by a consortium of federal agencies through a grant from the National Science Foundation (award number SES-0453930). Any opinions, findings, conclusions, or recommendations expressed in this publication are those of the author(s) and do not necessarily reflect the view of the organizations or agencies that provided support for this project.

International Standard Book Number-13: 978-0-309-14173-4
International Standard Book Number-10: 0-309-14173-7

Additional copies of this report are available from the National Academies Press, 500 Fifth Street, N.W., Lockbox 285, Washington, DC 20055; (800) 624-6242 or (202) 334-3313 (in the Washington metropolitan area); Internet, http://www.nap.edu.

Printed in the United States of America

Suggested citation: National Research Council. (2009). *Reengineering the Survey of Income and Program Participation.* Panel on the Census Bureau's Reengineered Survey of Income and Program Participation, Constance F. Citro and John Karl Scholz, Editors. Committee on National Statistics, Division of Behavioral and Social Sciences and Education. Washington, DC: The National Academies Press.

THE NATIONAL ACADEMIES

Advisers to the Nation on Science, Engineering, and Medicine

The **National Academy of Sciences** is a private, nonprofit, self-perpetuating society of distinguished scholars engaged in scientific and engineering research, dedicated to the furtherance of science and technology and to their use for the general welfare. Upon the authority of the charter granted to it by the Congress in 1863, the Academy has a mandate that requires it to advise the federal government on scientific and technical matters. Dr. Ralph J. Cicerone is president of the National Academy of Sciences.

The **National Academy of Engineering** was established in 1964, under the charter of the National Academy of Sciences, as a parallel organization of outstanding engineers. It is autonomous in its administration and in the selection of its members, sharing with the National Academy of Sciences the responsibility for advising the federal government. The National Academy of Engineering also sponsors engineering programs aimed at meeting national needs, encourages education and research, and recognizes the superior achievements of engineers. Dr. Charles M. Vest is president of the National Academy of Engineering.

The **Institute of Medicine** was established in 1970 by the National Academy of Sciences to secure the services of eminent members of appropriate professions in the examination of policy matters pertaining to the health of the public. The Institute acts under the responsibility given to the National Academy of Sciences by its congressional charter to be an adviser to the federal government and, upon its own initiative, to identify issues of medical care, research, and education. Dr. Harvey V. Fineberg is president of the Institute of Medicine.

The **National Research Council** was organized by the National Academy of Sciences in 1916 to associate the broad community of science and technology with the Academy's purposes of furthering knowledge and advising the federal government. Functioning in accordance with general policies determined by the Academy, the Council has become the principal operating agency of both the National Academy of Sciences and the National Academy of Engineering in providing services to the government, the public, and the scientific and engineering communities. The Council is administered jointly by both Academies and the Institute of Medicine. Dr. Ralph J. Cicerone and Dr. Charles M. Vest are chair and vice chair, respectively, of the National Research Council.

www.national-academies.org

PANEL ON THE CENSUS BUREAU'S REENGINEERED SURVEY OF INCOME AND PROGRAM PARTICIPATION

JOHN KARL SCHOLZ (*Chair*), Department of Economics, University of Wisconsin–Madison
F. JAY BREIDT, Department of Statistics, Colorado State University
LEONARD E. BURMAN, Maxwell School, Syracuse University
JOHN L. CZAJKA, Mathematica Policy Research, Inc., Washington, DC
PETER GOTTSCHALK, Department of Economics, Boston College
RONALD T. HASKINS, Economic Studies Program, The Brookings Institution, Washington, DC
V. JOSEPH HOTZ, Department of Economics, Duke University
JOHN ICELAND, Department of Sociology, Pennsylvania State University
JEROME P. REITER, Department of Statistical Science, Duke University
DEANNA T. SCHEXNAYDER, Ray Marshall Center for the Study of Human Resources, LBJ School of Public Affairs, University of Texas at Austin
ROBERT F. SCHOENI, Institute for Social Research, University of Michigan–Ann Arbor
JENNIFER VAN HOOK, Department of Sociology, Pennsylvania State University

CONSTANCE F. CITRO, *Study Director*
PATRICIA RUGGLES, *Study Director* (through June 2008)
CARYN KUEBLER, *Associate Program Officer* (through March 2008)
AGNES GASKIN, *Administrative Assistant*
DOUGLAS SYLVESTER, *Consultant,* College of Law, Arizona State University

COMMITTEE ON NATIONAL STATISTICS
2008-2009

Acknowledgments

The Panel on the Census Bureau's Reengineered Survey of Income and Program Participation (SIPP) thanks the many people who contributed their time and expertise to the preparation of this report. We are most appreciative of their cooperation and assistance. We particularly thank the leadership responsible for SIPP at the U.S. Census Bureau, Howard Hogan, associate director for demographic programs, and David Johnson, chief of the Housing and Household Economic Statistics Division, who provided consistent and strong support for the work of the panel. David merits special recognition as the innovative, energetic driver of the SIPP reengineering effort. We also thank the many Census Bureau staff who made informative presentations to the panel and who responded to our many requests for information and explanation with alacrity and good will, including Patrick Benton, Michael Berning, Judith Eargle, James Farber, Jason Fields, James Gibbs, Denise Lewis, Tracy Mattingly, Nat McKee, Sally Obenski, Cotty Smith, Martha Stinson, and Deborah Wagner.

We thank other individuals who made presentations to the panel in their areas of expertise, including John Abowd, Cornell University; Robert F. Belli, University of Nebraska; Heather Boushey, Center for American Progress; Janet Holtzblatt, U.S. Treasury Department; Howard M. Iams, Social Security Administration; Julia Lane, National Science Foundation; Donald Oellerich, U.S. Department of Health and Human Services; Ralph Rector, U.S. Treasury Department; and Arloc Sherman, Center for Budget and Policy Priorities. We are particularly grateful to Douglas Sylvester and his colleagues, Kirk Bardin and Jessica Wann, at Arizona State University, who prepared a thorough-going assessment for the panel of statutes for the

50 states regarding the potential availability for use in a reengineered SIPP of state administrative records for the Medicaid, Temporary Assistance for Needy Families, other cash benefits, unemployment insurance, and workers' compensation programs.

The panel is indebted to its staff. Patricia Ruggles, United Nations Statistics Division, served as study director from fall 2006 through summer 2008 and drew on her extensive knowledge of SIPP and its uses for welfare program policy analysis in organizing panel meetings and drafting the panel's report. She was ably assisted by Caryn Kuebler, associate program officer for the panel through March 2008, and Agnes Gaskin, administrative assistant. The report was edited by Christine McShane. Connie Citro, director of the Committee on National Statistics, served as study director from summer 2008 and saw the panel's report through to completion. She went well beyond the standard for a high-quality study director. The panel gives thanks for her intellectual leadership, report writing, and enthusiasm for this project.

Most importantly, I am indebted to the panel members themselves. They were extremely hard working, providing draft text and comments on several rounds of drafts on a complex topic. Special kudos go to John Czajka, who drafted the appendix on the quality of SIPP data and contributed materially to the data quality-related sections of Chapters 2 and 4 and to many other aspects of the report. Jay Breidt and Jerry Reiter made important contributions to the discussion of imputation methods and approaches for confidentiality protection in Chapter 3; Deanna Schexnayder provided informative material on arrangements for accessing state administrative records for welfare program research; and Jennifer Van Hook contributed material on immigration data and SIPP accessibility.

This report has been reviewed in draft form by individuals chosen for their diverse perspectives and technical expertise, in accordance with procedures approved by the Report Review Committee of the National Research Council (NRC). The purpose of this independent review is to provide candid and critical comments that will assist the institution in making its published report as sound as possible and to ensure that the report meets institutional standards for objectivity, evidence, and responsiveness to the study charge. The review comments and draft manuscript remain confidential to protect the integrity of the deliberative process. We wish to thank the following individuals for their review of this report: Robert F. Belli, Department of Psychology, University of Nebraska, Lincoln; Heather Boushey, Economic Policy, Center for American Progress; Jacob J. Feldman, Center for Health Research and Policy, Social and Scientific Systems; Howard M. Iams, Office of Retirement and Disability Policy/Office of Research Evaluation, and Statistics, Social Security Administration; Pamela Loprest, Income and Benefits Policy, The Urban Institute; Allen Schirm, Mathematica Policy Research;

Timothy M. Smeeding, Institute for Research on Poverty, Robert M. La Follette School of Public Affairs, University of Wisconsin–Madison; and Frank P. Stafford, Population Studies Center and Survey Research Center, University of Michigan.

Although the reviewers listed above have provided many constructive comments and suggestions, they were not asked to endorse the conclusions or recommendations nor did they see the final draft of the report before its release. The review of this report was overseen by Johanna T. Dwyer, Tufts University School of Medicine and Friedman School of Nutrition Science and Policy, Tufts Medical Center. Appointed by the NRC's Report Review Committee, she was responsible for making certain that an independent examination of this report was carried out in accordance with institutional procedures and that all review comments were carefully considered. Responsibility for the final content of this report rests entirely with the authoring panel and the institution.

Finally, we recognize the many federal agencies that support the Committee on National Statistics directly and through a grant from the National Science Foundation. Without their support and their commitment to improving the national statistical system, the committee work that is the basis of this report would not have been possible.

John Karl Scholz, *Chair*
Panel on the Census Bureau's
Reengineered Survey of Income and
Program Participation

Contents

Summary

The Survey of Income and Program Participation (SIPP) began 25 years ago to fill gaps in the available information on the short-term dynamics of income, household composition, employment, and eligibility for and participation in government assistance programs experienced by families in America. At present, SIPP follows samples of household members (panels) for 3-4 years, interviewing them every 4 months in order to obtain as accurate monthly information as possible and starting up a new panel when a previous panel ends.

Beginning in 2006, the Census Bureau embarked on a program to reengineer SIPP to reduce its costs and improve data quality and timeliness to the extent possible by such means as making greater use of administrative records, moving to annual interviews in which event history calendars would be used to ascertain monthly information, and modernizing the SIPP data collection and processing systems.[1] The bureau also requested the Committee on National Statistics of the National Academies to establish a study panel to address specific aspects of the reengineering program. The panel was asked to consider the advantages and disadvantages of strategies for linking administrative records and survey data, taking account of the accessibility of relevant administrative records, the operational feasibility of linking, the quality and usefulness of the linked data, and the ability to provide access to the linked data while protecting the confidentiality of

---

[1]Event history calendars are customized calendars that show the reference period, such as a year, and contain timelines for different domains, such as residence history, household composition history, work history, and other areas, that might aid a respondent's memory.

individual respondents. The panel also was charged to consider alternative uses of administrative records for a reengineered SIPP that do not require actual data linking (for example, to evaluate SIPP data quality). In addition, the panel could consider aspects of the reengineered SIPP survey with regard to interview periodicity, mode of data collection, and sample source and size.

The panel addressed the charge by first examining the history of SIPP to inform its deliberations about the survey's purpose, value, strengths, and weaknesses (Chapter 2). We then reviewed alternative uses of administrative records in a reengineered SIPP (Chapter 3) and, finally, considered innovations in SIPP design and data collection, including the proposed use of annual interviews with an event history calendar (Chapter 4). The panel's conclusions and recommendations from each chapter follow.

HISTORICAL PERSPECTIVE

Conclusion 2-1: The Survey of Income and Program Participation is a unique source of information for a representative sample of household members on the intrayear dynamics of income, employment, and program eligibility and participation, together with related demographic and socioeconomic characteristics. This information remains as vital today for evaluating and improving government programs addressed to social and economic needs of the U.S. population as it did when the survey began 25 years ago.

Conclusion 2-2: The Survey of Income and Program Participation's (SIPP) history of forward movement followed by setbacks has contributed to the survey's falling short of its original promise with regard to timeliness, usability, and maintenance of data quality. With the Census Bureau's planned SIPP reengineering program, there is an opportunity to put the survey on a much firmer foundation for the future. It is essential that the Census Bureau's program to reengineer SIPP address its problems and retain and build on its unique value and strengths.

No survey can be all things to all users. In reengineering SIPP, the focus should be on improving the content and design features of the survey that make possible its unique contribution.

Recommendation 2-1: To guide the design of a reengineered Survey of Income and Program Participation, the Census Bureau should consider the primary goal of the survey to be to provide data for policy analysis and research on the short-run (intrayear) dynamics of economic well-being for families and households, including employment, earnings, other income, and program eligibility and participation.

Recommendation 2-2: The Census Bureau's reengineering program for the Survey of Income and Program Participation should explicitly evaluate each proposed innovative feature, such as the use of administrative records or an event history calendar, on the extent to which a feature contributes to the survey's ability to measure short-term changes in economic well-being with improved quality and timeliness.

THE ROLE OF ADMINISTRATIVE RECORDS IN A REENGINEERED SIPP

Conclusion 3-1: In reengineering the Survey of Income and Program Participation (SIPP) to provide policy-relevant information on the short-run dynamics of economic well-being for families and households, the Census Bureau must continue to use survey interviews as the primary data collection vehicle. Administrative records from federal and state agencies cannot replace SIPP, primarily because they do not provide information on people who are eligible for—but do not participate in—government assistance programs and, more generally, because they do not provide all of the detail that is needed for SIPP to serve its primary goal. Many records are also difficult to acquire and use because of legal restrictions on data sharing, and some of the information they contain may be erroneous. Nonetheless, information from administrative records that is relevant to SIPP and likely to improve the quality of SIPP reports of program participation and income receipt in particular can and should be used in a reengineered SIPP.

Conclusion 3-2: The Census Bureau has made excellent progress with the Statistical Administrative Records System and related systems, such as the person validation system, in building the infrastructure to support widespread use of administrative records in its household survey programs. The bureau's administrative records program, both now and in the future as it adds new sets of records and analysis capabilities, will be an important resource for applications of administrative records in a reengineered Survey of Income and Program Participation.

Acquisition of Records

Conclusion 3-3: Many relevant federal administrative records are readily available to the Census Bureau for use in a reengineered Survey of Income and Program Participation (SIPP). However, most state administrative data are not available for use in a reengineered SIPP at this time and could be difficult to obtain.

Recommendation 3-1: The Census Bureau should seek to acquire additional federal records that are relevant to the Survey of Income and Program Participation, which could include records from the U.S. Department of Veterans Affairs and the Office of Child Support Enforcement.

Recommendation 3-2: The Census Bureau, in close consultation with data users, should develop a strategy for acquiring selected state administrative records, recognizing that it will be costly and probably unfeasible to acquire all relevant records from all or even most states. The bureau's acquisition strategy should be guided by such criteria as the importance of the income source for lower income households, particularly in times of economic distress, and the relative ease of acquiring the records. Unemployment insurance benefit records should be a high priority for the Census Bureau to acquire on both of these counts, and the bureau should investigate whether it is possible to acquire these records from the National Directory of New Hires, which would eliminate the need to negotiate with individual states.

Indirect Uses of Records

Conclusion 3-4: Indirect uses of administrative records are those uses, such as evaluation of data quality and improvement of imputation models for missing data, in which the administrative data are never recorded on survey records. They are advantageous for a reengineered Survey of Income and Program Participation (SIPP) in that they should have little or no adverse effects on timeliness or the needed level of confidentiality protection of SIPP data products.

Recommendation 3-3: The Census Bureau, in close cooperation with knowledgeable staff from program agencies, should conduct regular, frequent assessments of Survey of Income and Program Participation (SIPP) data quality by comparison with aggregate counts of recipients and income and benefit amounts from appropriate administrative records. When feasible, the bureau should also evaluate reporting errors for income sources—both underreporting and overreporting—by exact-match studies that link SIPP records with the corresponding administrative records. The Census Bureau should use the results of aggregate and individual-level comparisons to identify priority areas for improving SIPP data quality.

Recommendation 3-4: The Census Bureau should move to replace hot-deck imputation routines for missing data in the Survey of Income and Program Participation with modern model-based imputations, implemented multiple times to permit estimating the variability due to imputation. Imputation models for income and program participation should make use of program eligibility criteria and characteristics of beneficiaries from administrative records so that the imputed values reflect as closely as possible what is known about the beneficiary population. Before implementation, new imputation models should be evaluated to establish their superiority to the imputation routines they are to replace.

Recommendation 3-5: The Census Bureau should request the Statistical and Science Policy Office in the U.S. Office of Management and Budget to establish an interagency working group on uses of administrative records in the Survey of Income and Program Participation (SIPP).[2] The group would include technical staff from relevant agencies who have deep knowledge of assistance programs and income sources along with Census Bureau SIPP staff. The group would facilitate regular comparisons of SIPP data with administrative records counts of income recipients and amounts (see Recommendation 3-3) and advise the Census Bureau on priorities for acquiring additional federal and selected state administrative records, how best to tailor imputation models for different sources of income and program benefits, and other matters related to the most effective ways to use administrative records in SIPP. The Census Bureau should regularly report on its progress in implementing priority actions identified by the group.

Direct Uses of Records

Conclusion 3-5: Direct uses of administrative records in a reengineered Survey of Income and Program Participation (SIPP), which include substituting administrative values for missing survey responses, adjusting survey responses for net underreporting, using administrative values instead of asking survey questions, and appending additional administrative data, potentially offer significant improvements in the quality of SIPP data on income and program participation. They also raise significant concerns about increased risks of disclosure and delays in the release of SIPP data products.

Recommendation 3-6: In the near term, the Census Bureau should give priority to indirect uses of administrative records in a reengineered Survey of Income and Program Participation (SIPP). At the same time and working closely with data users and agencies with custody of relevant administrative records, the bureau should identify feasible direct uses of administrative records in SIPP to be implemented in the medium and

[2]See Recommendation 4-5 regarding an advisory group of outside researchers and policy analysts.

longer terms. Social Security and Supplemental Security Income benefit records, which are available to the Census Bureau on a timely basis, are prime candidates for research and development on ways to use the administrative values directly—either to adjust survey responses for categories of beneficiaries or to replace survey questions (which would reduce respondent burden)—in ways that protect confidentiality.

Recommendation 3-7: When considering the addition to the Survey of Income and Program Participation (SIPP) of administrative records values for variables that have never been ascertained in the survey itself, the Census Bureau should ensure that the benefits from the added variables are worth the costs, such as additional steps to protect confidentiality. The bureau should consult closely with users to be sure that the added variables are central to SIPP's purpose to provide information on the short-run dynamics of economic well-being and that their inclusion does not compromise the ability to release public-use microdata files that accurately represent the survey data.

Confidentiality Protection and Data Access

Conclusion 3-6: Multiple strategies for confidentiality protection and data access are necessary for a survey as rich in data as the Survey of Income and Program Participation. Public-use microdata files, which are available on a timely basis and in which confidentiality protection techniques do not unduly distort the relationships in the data, are the preferred mode of data release. Some uses may require access to confidential data that at present can be provided only at one of the Census Bureau's Research Data Centers.

Recommendation 3-8: The Census Bureau should develop confidentiality protection techniques and restricted access modes for the Survey of Income and Program Participation (SIPP) that are as user-friendly as possible, consistent with the bureau's duty to minimize disclosure risk. In this regard, the bureau should develop partial synthesis techniques for SIPP public-use microdata files that, based on evaluation results, are found to preserve the research utility of the information. For SIPP data that cannot be publicly released, the Census Bureau should give high priority to developing a secure remote access system that does not require visiting a Research Data Center to use the information. The bureau should also deposit SIPP files of linked survey and administrative records data (with identifiers removed) at all Research Data Centers in order to expand the opportunities for research that contributes to scientific knowledge and informed public policy.

INNOVATION IN DESIGN AND DATA COLLECTION

Event History Calendar Approach

Conclusion 4-1: The Survey of Income and Program Participation (SIPP) is the only national survey that provides information on the short-term dynamics of employment, income, program participation, and other family characteristics, and its monthly time frame is essential for many applications. The Census Bureau's plans to move SIPP to an annual survey, filling in intrayear dynamics using event history calendars, potentially affects—perhaps positively, perhaps negatively—SIPP's single most important feature.

Conclusion 4-2: The panel is not aware of conclusive evidence that a 12-month event history calendar (EHC) framework is capable (or not) of generating accurate monthly information on income, program participation, and other topics that are covered in the Survey of Income and Program Participation (SIPP). The lack of evidence about the ability of an EHC to collect monthly data places considerable pressure on the Census Bureau, not only to design an effective pretesting program for the EHC methodology, but also to make its survey reengineering plans for SIPP sufficiently flexible so that it can modify its plans if the pretesting reveals unanticipated, negative evidence on the likely success of the proposed methodology in providing high-quality monthly information.

Conclusion 4-3: Understanding transitions at the seam between interviews in a reengineered Survey of Income and Program Participation (SIPP) using the event history calendar approach will require data from at least two annual interviews. Moreover, not enough is yet known about the factors driving seam bias in the traditional SIPP.

Conclusion 4-4: A parallel traditional Survey of Income and Program Participation (SIPP) panel that provides 2 or more years of data is a necessary component of a thorough evaluation of the reengineered SIPP using the event history approach. The recently completed paper test is of limited value for this purpose. The Census Bureau's planned electronic prototype test is promising but, as a single test, is unlikely to provide conclusive findings.

Recommendation 4-1: The Census Bureau should engage in a major program of experimentation and evaluation of the event history approach for developing suitable data on the short-run dynamics of household composition, income, employment, and program participation from a reengineered Survey of Income and Program Participation (SIPP). The details of the

Census Bureau's plans should be disseminated to SIPP stakeholders for comment and suggestions for improvement. If the experimental results indicate that the quality of data on income and program dynamics is significantly worse under the event history calendar approach than in the traditional SIPP, the Census Bureau should return to a more frequent interview schedule, say, every 6 months, devise other methods to improve data on short-run dynamics, or revert to the traditional SIPP with 4-month interviews using standard questionnaires.

Recommendation 4-2: To ensure not only adequate evaluation of a reengineered Survey of Income and Program Participation (SIPP), but also a bridge between data collected under the new and old methods, the Census Bureau should conduct traditional and reengineered SIPP panels to provide at least 2 years of comparable data. If the new design works, then the parallel traditional panel provides a bridge. If the new design does not work, then the parallel panel provides a backup for the continued collection of SIPP data while the new design is modified as appropriate.

Recommendation 4-3: Because the reengineered Survey of Income and Program Participation (SIPP) should be compared with the first year of a traditional SIPP panel in order to minimize attrition bias, the Census Bureau should begin a new traditional SIPP panel in February 2012. If the costs of fielding two concurrent national longitudinal surveys appear prohibitive, the 2012 traditional SIPP panel could be smaller than previous SIPP panels without substantially diminishing its scientific value.

Length and Frequency of Interviews and Panels

Conclusion 4-5: Design features for a reengineered Survey of Income and Program Participation (SIPP) that are important to evaluate in terms of their effects on respondent burden, survey costs, data quality, and operational complexity include the length and frequency of interviews, the length of panels, and whether successive panels overlap. With regard to interviews, there is no evidence that a 12-month event history calendar strikes the optimal balance between respondent burden, costs, and data quality in comparison to the traditional SIPP design of 4-month interviews. With regard to panels, there is evidence that nonoverlapping panels have adverse effects on cross-sectional estimates of trends over time, yet they are advantageous in terms of larger sample sizes per panel and operational feasibility.

Recommendation 4-4: The Census Bureau should study the tradeoffs in survey quality and respondent burden in comparison to survey costs between longer but less frequent event history-based interviews in a reengineered Survey of Income and Program Participation (SIPP) and more frequent interviews in the traditional SIPP. The Census Bureau's research and evaluation program for SIPP should also improve understanding of panel bias and how it grows over time. Because overlapping panels remain the best way to document the extent of panel bias across the full range of variables collected in SIPP, they should be on the research agenda for possible implementation at a future time. Due to technical demands and capacity issues that arise in launching the reengineered SIPP, the initial design plans should not include overlapping panels.

Content

Conclusion 4-6: The Census Bureau has done an exemplary job in reaching out to the Survey of Income and Program Participation user community with "content matrices" and other efforts to identify critical portions of the core questionnaire and topical modules for data users.

Recommendation 4-5: The Census Bureau should expand the scope of the reconstituted Survey of Income and Program Participation (SIPP) Working Group or establish a new SIPP advisory group with members from academic institutions and policy research organizations that would meet periodically to assist the Census Bureau in its efforts to continually improve the quality and relevance of the SIPP survey content. This group, which could include government members from the recommended interagency working group on uses of administrative records in SIPP (see Recommendation 3-5), would review the Census Bureau's use of cognitive and other methods to evaluate and improve survey question wording and improve response rates (or, when that is not possible, either dropping the question or seeking an alternate data source); assist in benchmarking survey responses against external, reliable sources; and advise the bureau on ways to improve imputation and editing procedures. The group would provide a sounding board for the Census Bureau's plans to develop appropriate survey content in a reengineered SIPP and advise the bureau on appropriate modifications to survey content as policy developments occur, such as health care and immigration reform.

Timeliness

Conclusion 4-7: The release of Survey of Income and Program Participation (SIPP) data is often not timely. Data from the 2004 SIPP panel were generally released more than 2 years after being collected. Other panel surveys have more timely data release, often within a year of data collection, which enhances their usefulness to external users.

Recommendation 4-6: The Census Bureau should release Survey of Income and Program Participation data within 1 year of data collection.

Management and Budget

Conclusion 4-8: Unlike other surveys of people and households that the Census Bureau conducts, the Survey of Income and Program Participation (SIPP) does not have a government client outside the Census Bureau or a federally mandated set of reports that are based on the survey. Not having an external client, such as the Bureau of Labor Statistics (which has a collaborative and financial stake in the monthly Current Population Survey), or a set of regular reporting requirements, as with the decennial census and the American Community Survey, has contributed to setbacks in the development of SIPP. The value of the survey has also been diminished over its history by sample cutbacks necessitated by cutbacks in funding.

We agree with an earlier Committee on National Statistics panel (National Research Council, 1993) that SIPP would benefit from a project director with full management and budget authority for design, evaluation, and operations. The budget should always include adequate research and development funding, since SIPP is a major ongoing survey that requires regular evaluation and improvement.

1

Introduction

The Survey of Income and Program Participation (SIPP) is a continuing program of the U.S. Census Bureau, which began interviewing households for the survey in late 1983 and is planning to introduce a major redesign and reengineering of the survey beginning in 2013. Under its current design, in which members of sampled households (panels) are interviewed every 4 months for 3 or 4 years, SIPP provides vital information for planning, evaluation, and improvement of government programs intended to address social and economic needs of the U.S. population. Uniquely among surveys, SIPP not only provides detailed information on incomes by source for a representative sample of U.S. households, but also tracks changes in program eligibility and participation for the members of those households as their incomes and other circumstances change. Understanding these changes is essential for government social welfare program planning and evaluation.

To make the survey more cost-effective while improving to the extent possible the quality and timeliness of the data, the Census Bureau began a research and development program in 2006 to assess new ways to collect, process, and disseminate the data. As one component of its program to reengineer SIPP, the Census Bureau requested the Committee on National Statistics of the National Academies to convene a panel to study technical issues of using administrative records as part of SIPP. The panel was charged to consider the advantages and disadvantages of strategies for linking administrative records and survey data, taking account of the accessibility of relevant administrative records, the operational feasibility of linking, the quality and usefulness of the linked data, and the ability to provide access to the linked data while protecting the confidentiality of individual respondents. The panel was also charged to consider alternative uses of administrative records for a reengineered SIPP that do not require actual data linking (for example, to evaluate SIPP data quality). In addition, the panel could consider aspects of the reengineered SIPP survey with regard to interview periodicity, mode of data collection, and sample source and size.

SIPP IN BRIEF

Before SIPP was initiated in the early 1980s, government experts and scholars agreed that better data on incomes and program participation were needed in order to assess and redesign social programs (see National Research Council, 1993:26-28). The major source of such data, the Current Population Survey (CPS), provided only limited information on family incomes and participation in government programs. This information was inadequate, not only because the CPS income reporting period (the previous calendar year) did not match the income reporting period for programs (the previous month in many instances), but also because the data did not allow researchers to track individuals and families over time. Experience in administering such programs as unemployment insurance and food stamps indicated that at least some program participants faced frequent changes in employment, earnings, and income, and that these changes were often associated with changes in program eligibility and participation that were important to understand.

A new survey that followed the same individuals over time, recording as many of these changes in income and program participation as possible, was therefore needed. To fill this gap, the Office of the Assistant Secretary for Planning and Evaluation and the Social Security Administration in what was then the U.S. Department of Health, Education and Welfare worked with the Census Bureau and outside researchers over a period of years to conceptualize, design, and field test survey questions and methods for a new survey. SIPP was the result.

Interviews for the first SIPP panel of households began in fall 1983, and, with a few exceptions, a SIPP panel has been in the field every year since then. Each panel consists of the members of a representative sample of households (ranging in size from 12,000 to 51,000 households at the start of a panel), who are interviewed every 4 months about their income, employment, family relationships, and program participation for each of the 4 months preceding the interview. Most panels have continued for 2-4 years. In the early years of the survey, SIPP interviewers conducted in-person interviews of sample members using paper and pencil questionnaires. At present, SIPP interviewers use computer-assisted personal interviewing (CAPI) for the first two interview waves and computer-assisted telephone interviewing (CATI) for all subsequent waves.

Originally, a new SIPP panel began every year; in a redesign introduced in 1996, a new panel begins every 3 or 4 years following the conclusion of the previous panel. The selection of the sample for each panel is a complex procedure that results in a probability sample of the U.S. population (excluding only inmates of institutions and armed forces members living on base without their families), with oversampling of low-income households based on their census characteristics. The sample includes cases in every state and the District of Columbia, although SIPP currently can support reliable state-level estimates for only 14 states.

In addition to the core data asked in every interview wave, SIPP includes topical modules, which are sets of questions asked one or more times of each panel on a wide range of subjects. Responses to topical module questions on child care arrangements, child well-being, marital, fertility, and employment history, pension rights, asset holdings, and other subjects broaden and deepen the analyses that can be conducted with SIPP data on important social and economic welfare issues of public policy concern.

SIPP'S UNIQUE CONTRIBUTION

The unique feature of SIPP is its capacity to measure dynamics *in the short run*. Monthly data on incomes and demographic characteristics of households allow analysts to study intrayear transitions in marital status, poverty, employment, health insurance coverage, and eligibility for and participation in a wide range of government programs. These kinds of analyses are not possible with other nationally representative data sets, which require respondents to recall income amounts, program participation, and other characteristics for an entire year, not just 4 months, as in SIPP.

The monthly time frame is critical given that eligibility for many public programs is assessed on a monthly basis and that people may have short spells of both program eligibility and participation. While administrative data can be used to look at dynamic patterns of participation in a single program, only SIPP, which includes both participants and nonparticipants in a wide range of programs, can be used to examine dynamic patterns of eligibility—and of participation contingent on eligibility—in single and multiple programs at the same or different time periods. For example, in one of the first such analyses, considering Social Security, Supplemental Security Income, public assistance, and food stamps, Doyle and Long (1988) estimated that 17 percent of people in the first month of the 1984 SIPP panel participated in a single program, and another 6 percent participated in more than one of these programs. Moreover, during the next 11 months, about 6 percent of the initial program recipients experienced at least one transition to a different program combination or ended their participation.

Choosing the right time interval for a specific policy analysis of program participation and eligibility can have important consequences. The free and reduced-price school lunch program offers an example. A report of the Food and Nutrition Service of the U.S. Department of Agriculture (USDA) suggested that the number of children certified for free meals in 1999 was 27 percent greater than the number who appeared to be eligible, indicating extensive "overcertification" in the school meals program (Food and Nutrition Service, 1999). Results like this contributed to the Improper Payments Information Act of 2002, which requires that various federal agencies identify and reduce erroneous payments in their programs. The USDA is one of the agencies the act targeted.

The CPS Annual Social and Economic Supplement, the source of the data used for the Food and Nutrition Service's overcertification results, collects only annual data on income. Annual income necessarily smoothes month-to-month variation in income—yet it is *monthly* income that statutorily affects eligibility. Parents or guardians self-report household income for the calendar month prior to the application for free or reduced-price school meals. Income must be equal to or less than 130 percent of the poverty line or the household must receive food stamps or Temporary Assistance for Needy Families benefits for the children to be eligible for free school lunches. Using data from SIPP, which allowed them to calculate eligibility based on information that mirrors the statutes governing program eligibility, Dahl and Scholz (2005) report that participation in the school lunch program (as a fraction of eligible children) is 77 percent, far lower than USDA's CPS–based estimate of 127 percent. The Dahl and Scholz estimate covers free meals for the period from 1993 through 2003. The differences between the two studies' results are large, and estimates of program cost, take-up (i.e., the percentage of eligible people who apply for and receive program benefits), and the consequences of altering program rules simply cannot be made without accurate monthly data on the eligible population.

The fact that the SIPP monthly data are obtained from a panel survey—in which sample members are followed over time, rather than from a cross-sectional survey collecting retrospective monthly information—is important not only for longitudinal uses of the data, but also for cross-sectional, point-in-time analyses. Households and families are dynamic, experiencing such events as the birth of a child or the loss of a parent due to death or divorce, and these changes affect the income and other resources available to household members during a year or other time period. SIPP can capture these kinds of changes, which cross-sectional surveys cannot. Research on poverty has shown the importance of the panel feature of SIPP for cross-sectional analyses. Annual poverty rates estimated from SIPP are consistently lower than those from the CPS (see, for example, Lamas, Tin, and Eargle, 1994). These differences are due to several factors, including that income reporting for the low-income population is more complete in SIPP with its 4-month interviews than in the annual CPS retrospective interviews. Another factor is that the CPS poverty rates are based on the characteristics of a family at a point in time and do not capture the income that may have been available to family members during the year from people who were in the family only part of the year.

REENGINEERING

Despite a record of providing invaluable data for important research and policy studies, SIPP has experienced many ups and downs over its 25 years of existence (see Chapter 2). Periodically, budget cuts have necessitated cuts in sample size and the length of panels. Some problems have plagued the survey from its inception, such as late delivery of data files to users, complex file structures, and inadequate documentation, which make it difficult for users to work with the data. In addition, growing rates of attrition of sample members over the life of a panel (both at the first interview wave and in subsequent waves), underreporting of program recipients and benefit amounts when compared with administrative records, and other factors have led to concerns about the quality of the data.

The Census Bureau recognized the need to reengineer the outmoded SIPP data processing system, which contributed to delays in data release, and to address other problems. But the budget climate was not favorable to making the needed investment in the survey. In January 2006, when required by the Office of Management and Budget to absorb a significant budget cut, the bureau decided to discontinue SIPP. Congress, however, encouraged by an outpouring of support for the survey from data users (see Chapter 2), appropriated funds not only to continue SIPP in its current form, but also to reengineer the survey to be more timely and cost-effective in the future. The Census Bureau is well along on its reengineering agenda, which includes testing an event history calendar approach to collecting the core data. This approach may permit interviews to be scaled back from three interviews to just one per year. Another aspect of the reengineering agenda is the search for cost-effective uses of administrative records, such as federal and state tax and transfer program records, to assess the quality of responses to questions on the SIPP interview or to supplement the survey with additional information.

ORGANIZATION OF THE REPORT

This report with its conclusions and recommendations is organized into four chapters and two appendixes. Chapter 2 fleshes out the history of SIPP from the early days of its conceptualization through the present period of redesign. The chapter describes the strengths of SIPP and the challenges it presents to users in terms of data quality, timeliness, and complexity of data files.

Chapter 3 discusses the possible roles for federal and state administrative records in a reengineered SIPP, which include their use to evaluate the quality of survey reports, improve imputations used to provide values for missing responses, correct survey responses for misreporting, and replace survey questions. The chapter considers the costs and benefits of each major use of administrative records and both short-term and longer term goals for making the best use of records for a reengineered SIPP. An important consideration for expanding the role of administrative records in SIPP concerns the consequences for access to microdata for research and policy analysis while protecting the confidentiality of individual responses. The extent to which confidentiality protection becomes more difficult than with the current design depends heavily on the specific roles that are identified for administrative records in a new SIPP design.

Chapter 4 discusses proposed innovations in design and data collection for SIPP that may interact with proposed uses of administrative records. The chapter focuses on the planned use of an event history calendar to collect intrayear data of high quality at less frequent intervals than under the current design. It identifies potential strengths and weaknesses of that approach in comparison with the current SIPP design and outlines a comprehensive set of evaluations for understanding the consequences of adopting an event history calendar approach. It also addresses related considerations of the length and number of interviews and the length and overlap of panels, along with issues of data content, timeliness, and budget for a reengineered SIPP. It briefly addresses the SIPP sample size and design.

The appendixes provide additional information on SIPP data quality and the backgrounds of panel members and staff.

2

SIPP's History, Strengths, and Weaknesses

This chapter briefly reviews the history of the Survey of Income and Program Participation (SIPP) from the perspective of its original goals and summarizes plans for reengineering the survey.[1] It describes SIPP's strengths under its current design—strengths that a new design needs to maintain. It also describes SIPP's weaknesses, which a new design needs to ameliorate to the extent possible. Conclusions and recommendations are provided at the end of the chapter.

HISTORY

From its earliest days to the present, SIPP has exhibited a pattern of a forward movement, followed by a setback, followed by another forward movement, another setback, and so on. This pattern has adversely affected the usefulness, quality, and cost-effectiveness of the data at various times. Yet, overall, the survey has shown a marked resilience and has earned the support of users who find the SIPP data indispensable for important kinds of policy analysis and research.

[1]Principal sources of information for this chapter include National Research Council (1993); Citro (2007); and presentations by David Johnson, chief of the Census Bureau's Housing and Household Economic Statistics Division, in 2006 and 2007 (available at http://www.census.gov/sipp/dews.html; see also http://www.census.gov/sipp/).

Origins and Goals

The origins of SIPP date to the late 1960s, when policy makers trying to implement antipoverty programs under the War on Poverty expressed dissatisfaction with the quality and detail of data on income and welfare program participation available from the Current Population Survey (CPS) March Income Supplement.[2] In 1975 the then U.S. Department of Health, Education and Welfare (HEW) established the Income Survey Development Program (ISDP). Responsibility for designing and analyzing a new survey was shared between the Office of the Assistant Secretary for Planning and Evaluation (ASPE) and the Social Security Administration (SSA), both in HEW at the time. The U.S. Census Bureau was charged with collecting the survey data.

The ISDP conducted experiments at five test sites in 1977. Next, a 1978 ISDP research panel followed members of about 2,340 original sample households through several interviews. Finally, a 1979 ISDP research panel followed members of about 9,500 original sample households over 6 interviews every 3 months, for a total of 18 months. The interviews asked about monthly employment, income, and program participation; asset income was ascertained once every 6 months.

At about the same time, when plans were well along to implement a new survey to be called SIPP, an interagency memorandum was drawn up in 1980 stating the survey's goals (see Kasprzyk, 1988). Signed by representatives of SSA, ASPE, and the Census Bureau, the memorandum stipulated that SIPP's goals were to

1. extend the scope and precision of policy analyses for a wide range of federal and state tax and social welfare programs;
2. improve current estimates of income and income change, including annual and subannual estimates, by source of income; and
3. broadly assess the economic well-being of the population.

First Crisis

SIPP's first crisis occurred at the moment when it was officially supposed to begin. The transition from the ISDP to SIPP was scheduled for 1981, with operational control of the survey transferred from ASPE to SSA. While ASPE and the Census Bureau were to remain as partners in

[2]The CPS March Income supplement was renamed the CPS Annual Social and Economic Supplement (ASEC) when the sample for the supplement was expanded to improve the reliability of state estimates of children's health insurance coverage and some of the cases were interviewed in February and April (most cases are still interviewed in March). This sample expansion was first implemented in the 2002 CPS. Hereafter, we use "the CPS" when discussing income and program participation information from the supplement and "the monthly CPS" when discussing the core data on labor force participation that are collected every month.

the survey, the bulk of the funding was in the SSA budget. The election of Ronald Reagan as president, however, brought new policy priorities to the federal government. These new priorities caused the new administration and Congress to cancel SIPP.

In 1982 SIPP experienced the first of many last-minute reprieves. Bruce Chapman, the new director of the Census Bureau, convinced the White House to restore its funding. He argued that because SIPP would record more income than the CPS (based on the ISDP tests), it would produce a lower poverty rate, compared with the official poverty rate computed from the CPS. In restoring SIPP, full funding went to the Census Bureau, rather than being funneled through user agencies, such as ASPE and SSA, as originally planned.

The First Decade (1983-1993)

The first SIPP panel (the 1984 panel) began in October 1983. It originally included about 21,000 sample households, whose adult members age 15 and older the Census Bureau attempted to follow for 8 or 9 waves of interviews conducted every 4 months. However, 7 percent of the sample had to be dropped after Wave 4 because of budget cuts. Original sample members who moved within the United States were interviewed at their new address, unless they moved into an institution or became homeless. People in institutional settings and homeless people were not part of the sample, nor were people who moved outside the United States. Children under age 15 and adults who moved in with an original sample member after the first interview wave were included in the data collection so long as they resided with the original sample member.[3]

The SIPP design called for sample members to be interviewed every 4 months in order to increase the accuracy of answers to core questions on income amounts, participation in social programs, employment status, and health insurance coverage on a month-by-month basis compared with interviews at longer time intervals. Experiments conducted in the ISDP supported the use of 3-month interviews compared with 6-month interviews (Ycas and Lininger, 1983:28), and 4 months was a compromise. The core of each interview also included questions on key background characteristics, such as education, family composition, and ages of household members. In addition to the core questions, SIPP included one or more topical modules on important issues related to well-being and social policy. Questions in the topical modules, which covered a wide range of subjects (see Box 2-1), were asked only once or twice in a single panel.

[3]The 1993 SIPP panel followed children under age 15 even if they no longer resided with an original sample adult; however, the practice was abandoned in subsequent panels because so few such children were actually located.

BOX 2-1
Topical Modules in SIPP Panels, 1984-2004

Child care and support modules

- Child Care (once or twice in every panel)
- Child Support Agreements (once or twice in every panel beginning in 1985)
- Child Support Paid (2-4 times in 1996, 2001, 2004 panels)
- Informal Care-Giving (once in 2004 panel)
- Support for Nonhousehold Members (once or twice in every panel)
- Welfare History and Child Support (once in 1984 panel)

Disability and health care utilization modules

- Disability Status of Children (once or twice in 1985-1989 panels)
- Employer-Provided Health Benefits (once in 1996, 2001, 2004 panels)
- Functional Limitations and Disability (once or twice in 1990-1991 panels); separate modules for adults and children (once or twice in 1992, 1993, 1996, 2001, 2004 panels)
- Health and Disability (once in 1984 panel)
- Health Status and Utilization of Health Care Services (1-3 times in every panel beginning in 1985)
- Home Health Care (once in 1988-1989 panels)
- Long-Term Care (once or twice in 1985-1989 panels)
- Medical Expenses and Work Disability (once in 1987-1992 panels; 2-4 times in 1993, 1996, 2001, 2004 panels)
- Work Disability History (once early in every panel beginning in 1986)

Education modules

- Education and Training and Education and Work History (once each in 1984 panel)
- Education and Training History (once early in every panel beginning in 1986)
- School Enrollment and Financing (once or twice in every panel through 1996)

Employment modules

- Employment History (once early in every panel beginning in 1986)
- Home-Based Self-Employment/Size of Firm (once in 1992-1993 panels)
- Job Offers (once in 1985-1986 panels)
- Reasons for Not Working/Reservation Wage (once in 1984 panel)
- Time Spent Outside Workforce (once in 1990 panel)
- Work Expenses (once or twice in 1984-1987 panels; 2-4 times in 1996, 2001, 2004 panels)
- Work Schedule (once or twice in every panel beginning in 1987)

Family background modules
- Family Background (once in 1986-1988 panels)
- Fertility History (once early in every panel)
- Household Relationships (once early in every panel)
- Marital History (once early in every panel)
- Migration History (once early in every panel)

Financial modules
- Annual Income and Retirement Accounts (once or twice in every panel; 3 times in 1996 panel)
- Assets and Liabilities (once or twice in every panel; 3-4 times in 1996, 2001 panels)
- Housing Costs, Conditions, and Energy Usage (once in 1984 panel)
- Retirement Expectations and Pension Plan Coverage (once in most panels)
- Selected Financial Assets (once in selected panels)
- Shelter Costs and Energy Usage (once in 1986-1987 panels)
- Taxes (once or twice in every panel)

Program participation modules
- Real Estate Property and Vehicles (once or twice in most panels—for determining program eligibility)
- Real Estate, Shelter Costs, Dependent Care, and Vehicles (once or twice in selected panels—for determining program eligibility)
- Recipiency History (early in every panel beginning in 1986)
- Welfare Reform (once in 1996, 2001, 2004 panels)

Well-being modules
- Adult Well-Being (once in 1993, 1996, 2001 panels)
- Basic Needs (once in 1993 panel)
- Child Well-Being (1-3 times in 1993, 1996, 2001, 2004 panels)
- Extended Measures of Well-Being (once in 1991-1992 panels)

NOTE: Over the history of SIPP, the content of some topical modules changed with no change in title or the title changed with little change in content. Sometimes two topical modules with different titles have had similar content. There were no topical modules in Waves 9-12 of the 2004 panel. The actual questions are provided with the microdata technical documentation for the SIPP public-use files from the Census Bureau.

SOURCE: See http://www.census.gov/sipp/top_mod/top_mods_chart.html.

Building on the 4-month interval between interviews, a SIPP sample is divided into four equally sized rotation groups, which are interviewed in successive months. In addition to distributing the survey fieldwork uniformly over time, this rotation group structure ensures that the survey estimates for a given calendar month represent an average of responses given 1, 2, 3, and 4 months later.[4] Thus, any response bias associated with the reference month—for example, a decline in accuracy with distance from the interview—will affect all calendar months equally.

New SIPP panels began every February from 1985 through 1993. These panels were designed to overlap in time, so that samples from two different panels could be combined to provide representative cross-sectional estimates for a given year of the poverty rate and other characteristics, whereas a single panel would provide longitudinal information on intrayear transitions in employment, poverty status, and other characteristics for a sample of people followed over 2-3 years. However, because of lack of time and resources, the Census Bureau did not combine panels for analytical use, although it did provide factors to apply to the panel weights so that users could produce estimates from two overlapping panels.

The sample design for each panel was a multistage clustered probability sample of the population in the 50 states and the District of Columbia that excluded only inmates of institutions and those members of the armed forces living on base without their families. There was no oversampling of specific population groups in SIPP in the 1984-1993 panels, except that the 1990 panel included about 3,800 extra households continued from the 1989 panel, most of them selected because they were headed by blacks, Hispanics, or female single parents at the first wave of the 1989 panel.

Original sample sizes for the 1984 through 1993 panels ranged from 12,400 to 21,800 households, and the number of interview waves ranged from 3 to 10 (see Table 2-1, which also includes information for the 1996, 2001, and 2004 panels). In the early years of the survey, SIPP interviewers conducted in-person interviews of sample members using paper and pencil questionnaires. Telephone interviewing was tested in the 1986 panel (Gbur, Cantwell, and Petroni, 1990) and first used on a production basis in February 1992 in Wave 7 of the 1990 panel and Wave 4 of the 1991 panel. In the 1992 and 1993 panels, SIPP interviewers conducted in-person interviews for Waves 1, 2, and 6 and telephone interviews to the maximum extent possible for the other waves.

In the 1984, 1985, and 1986 panels, SIPP did not collect all of the information, such as shelter costs and medical expenses, that was necessary

[4]The rotation group design is incorporated into the SIPP survey weights as well. All cross-sectional and longitudinal weights are calculated so that each rotation group represents one-quarter of the population.

TABLE 2-1 Characteristics of SIPP Panels, 1984-2004: Number of Waves, Original Sample Size, Reduced Sample Size (if applicable), and Cumulative Sample Loss by Wave

Panel	No. of Waves	Original Sample Size	Reduced Sample Size (in Wave x)	Cumulative Sample Loss at Wave 1	3	6	9	12	Overall Loss (Final Wave)[d]
1984	9 (8)[a]	20,900	19,500 (5)	4.9	12.3	19.4	22.3	N.A.	22.3 (9)
1985	8 (7)[b]	14,300	13,500 (4)	6.7	13.2	19.7	N.A.	N.A.	20.8 (8)
1986	7 (6)[b]	12,400		7.3	15.2	20.0	N.A.	N.A.	20.7 (7)
1987	7	12,500		6.7	14.2	18.9	N.A.	N.A.	19.0 (7)
1988	6	12,700		7.5	14.7	18.3	N.A.	N.A.	18.3 (6)
1989	3	12,900		7.6	13.8	N.A.	N.A.	N.A.	13.8 (3)
1990[c]	8	19,800		7.3	14.4	20.2	N.A.	N.A.	21.0 (8)
1991	8	15,600		8.4	16.1	20.3	N.A.	N.A.	21.4 (8)
1992	10	21,600		9.3	16.4	21.6	26.2	N.A.	26.6 (10)
1993	9	21,800		8.9	16.2	22.2	26.9	N.A.	26.9 (9)
1996	12	40,200		8.4	17.8	27.4	32.8	35.5	35.5 (12)
2001	9	40,500	30,500 (2)	13.3	24.7	28.2	31.9	N.A.	31.9 (9)
2004	12	51,400	21,300 (9)	14.9	25.6	31.2	34.0	36.6	36.6 (12)

NOTES:

N.A. = Not applicable.

Original and reduced sample sizes are rounded to the nearest hundred households.

Original sample sizes are the number of households eligible to be interviewed at the start of Wave 1.

Reduced sample sizes are reductions in original sample sizes due to budget cuts (the wave in which the cut took effect is indicated in parentheses). Reduced sample sizes do not reflect reduction in sample sizes due to attrition; nor do they reflect growth (or decline) in sample sizes due to changes in household composition because original sample people moved out of or back into an original sample household.

Sample loss rates consist of cumulative noninterview rates adjusted for unobserved growth in the noninterviewed units (created by household splits after Wave 1). There are some differences in the calculation of sample loss between the 1984-1993 and 1996 panels, which allowed nonresponding households to drop out of a panel permanently if they missed a specific number of waves, and the 2001and 2004 panels, which kept all nonresponding households after Wave 1 in the sample.

[a]Two rotation groups in the 1984 panel received nine interview waves; the other two groups received eight waves—one group skipped Wave 2, and another group skipped Wave 8 in order to align the timing of collection of income tax information.

[b]One rotation group in each of the 1985 and 1986 panels received one fewer wave than the other three groups in order to collect income tax information in approximately the same time period.

[c]Sample loss rates are for the nationally representative portion of the sample; they exclude about 3,800 extra households headed by blacks, Hispanics, or single parents that were continued from the 1989 panel.

[d]The last wave of interviewing in a panel is indicated in parentheses.

SOURCE: Tabulations provided by Census Bureau staff.

to simulate eligibility for assistance programs, such as food stamps and Aid to Families with Dependent Children. Eligibility determination is essential to understand trends in program take-up rates—for example, an increase in the number of program participants could be due to an expansion of eligibility that did not alter the take-up rate, or to an increase in the take-up rate among already-eligible participants who decided to apply for benefits, or to both factors. In response to user requests, Wave 6 of the 1987 panel collected information on selected financial assets and medical expenses and disability that allowed eligibility simulations to begin. These modules were asked once in each of the 1988, 1990, 1991, and 1992 panels and twice in the 1993 panel. Beginning with Wave 7 of the 1993 panel, the eligibility modules were combined with the assets and liabilities module, and the combined modules were asked annually in the 1996 and subsequent panels.

Budget shortfalls necessitated a reduction in sample size in the 1984 and 1985 SIPP panels. Budget constraints also limited the sample size and number of interview waves in all panels initiated between 1985 and 1989 (see Table 2-1). These reductions and fluctuations in panel size and length made it difficult to plan for either fieldwork or analysis with confidence. Some of these panels were so short as to be effectively useless, except for cross-sectional analyses.

Consequently, the U.S. Office of Management and Budget (OMB) requested an evaluation of SIPP, and in 1990 the Census Bureau asked the Committee on National Statistics (CNSTAT) for a report on the future of SIPP (National Research Council, 1993).[5] Funding was also secured to boost somewhat the sample sizes and length of the 1990-1993 panels.

The Future of the Survey of Income and Program Participation, the CNSTAT panel's report, appeared in 1993 and contained a long list of recommendations for improving the content, design, and operation of the survey. After the CNSTAT panel report and an internal evaluation, SIPP underwent a major redesign that became effective with the 1996 panel. To maximize sample size for longitudinal analysis with a single panel and so reduce the need to combine panels for analysis purposes, as well as to reduce the burden on the field interviewers, the practice of introducing a new panel every year was dropped. The CNSTAT panel had recommended introducing new panels every 2 years and continuing them for 4 years, so that two panels would be in the field at any one time. The Census Bureau decided on a design that ran a panel through to completion before starting another panel.

[5]The full-scale study was preceded by an interim evaluation (National Research Council, 1989).

The Second Decade (1996-2006)

The 1996 panel began with 40,200 original sample households, of which about 36,800 (92 percent) were interviewed in Wave 1 and their members followed for as many waves as possible through 12 waves (4 years). Households in a high-poverty stratum based on their 1990 census characteristics were oversampled relative to other households. SIPP also kept track of original sample members who moved into institutions and resumed interviews with them when and if they rejoined the household population. An effort to field a new 4-year panel in 2000 was aborted because of the need for staff to devote full attention to the 2000 census. The 2001 panel began with about 40,500 households, of which about 35,100 were interviewed in Wave 1 (87 percent); the sample was reduced by 25 percent for budgetary reasons in Wave 2, and the members of the remaining sample households were followed for as many waves as possible through nine waves. The 2004 panel began with about 51,400 households, of which about 43,700 (85 percent) were interviewed in Wave 1; the members of these households were followed for as many as 12 waves, with a 58 percent sample reduction in the last four waves to free up resources to reengineer the survey (see "Reengineering (2006-Present)," below). Both the 2001 and 2004 panels oversampled the low-income population based on census characteristics (the 1990 census for the 2001 panel and the 2000 census for the 2004 panel).

The interviews from these various SIPP panels have yielded important data on a range of policy-related issues.[6] For example, Vaughan (2007) analyzed data from the first year of the 1996 SIPP panel to identify factors that would be likely to facilitate or impede the ability of participants in the Aid to Families with Dependent Children (AFDC) to make the transition from welfare to work subsequent to implementation of the Public Responsibility and Work Opportunity Reconciliation Act of 1996 (PRWORA, or welfare reform). He found that nearly half of AFDC recipients possessed two or more attributes (such as a disability) that impeded work in the period of transition to the new regime, in which work was the primary emphasis of the program. Only 30 percent of these participants held a job in 1996. In contrast, 41 percent of recipients possessed three or more attributes that facilitated work, and 68 percent of them held a job during 1996. A finding of note was that the age of the participants' children did not seem to represent a substantial barrier to work.

The Congressional Budget Office (CBO, 2003) analyzed data on health insurance coverage from the SIPP 1996 panel, the 1998 and 1999 Medical

[6]A recent estimate puts the number of publications based on SIPP at over 2,000 books, articles, reports, and other written products issued through 2006 (see http://www.census.gov/sipp/aboutbib.html).

Expenditure Panel Survey (MEPS), and the 1998 and 1999 CPS. From the SIPP and MEPS data, CBO estimated that 21 to 31 million people lacked coverage for an entire year in 1998 compared with the widely cited CPS estimate of about 44 million. CBO also estimated from SIPP and MEPS that about 41 million people lacked health insurance coverage at a specific point in time in 1998, while about 58 million lacked coverage at some point during the year. Looking at the duration of spells without coverage experienced by nonelderly people using 11 waves of the SIPP 1996 panel, CBO estimated that 45 percent of the uncovered spells that began between July 1996 and June 1997 lasted only 4 months, while 26 percent lasted 5-12 months, and 29 percent lasted more than a year. These different measures of health insurance coverage and estimates of duration of spells have important implications for the design of more effective health care coverage in the United States.

A companion Survey of Program Dynamics (SPD) is part of SIPP's history in its second decade. The SPD was mandated by PRWORA; it followed households that completed the 1992 and 1993 SIPP panels annually from 1997 through 2002. The SPD core instrument asked about employment, income, program participation, health insurance and utilization, child well-being, marital relationships, and parents' depression. SPD topical modules included a self-administered adolescent questionnaire asked in 1998 and 2001, additional child-related questions asked in 1999 and 2002, and residential histories of children asked in 2000. The SPD experienced some of the same problems as the main SIPP (see "Strengths and Weaknesses of SIPP Data" below), including not only data processing delays, but also high attrition rates until additional efforts and incentives were used to bring households back into the survey. (For more information, see http://www.census.gov/spd/overview.html; http://www.census.gov/spd/reports/pu02strp.html.)

Crisis in 2006

Like many events in the nation's capital, SIPP's most recent crisis resulted from a threatened cutoff of funds. In its *Budget for Fiscal Year 2007* (delivered to Congress in February 2006), the Bush administration planned for the Census Bureau to cut $40 million from its budget as part of a larger set of proposed reductions in domestic spending. Given the bureau's need to prepare for the 2010 decennial census, its choices were limited. Essentially, the cut could be accomplished either by taking pieces away from several different programs, making each of them less effective, or by eliminating one program and allowing the remainder to keep to their planned budgets and schedules.

This time, rather than continue a strategy of "death by a thousand cuts" in several programs, the Census Bureau decided that it would be preferable simply to drop one whole program. SIPP was chosen as the program to drop for a number of reasons. First, and perhaps most important, SIPP had no outside agency sponsor or legal mandate. The Census Bureau cannot unilaterally choose to terminate most of its surveys, because other departments and agencies depend on their output and frequently contribute to their budgets. The monthly CPS, for example, is sponsored by the Bureau of Labor Statistics (BLS), even though the data are collected by Census Bureau field staff. The monthly CPS is used to calculate monthly unemployment statistics, and the need to keep producing those statistics would prevent the Census Bureau from making major changes in the survey on its own initiative. Similarly, various economic surveys produce data used in computing the National Income and Product Accounts, the source of data on the gross domestic product. Changing any of these surveys would provoke protests from the Bureau of Economic Analysis, the Federal Reserve Board, and other agencies that use their data, and so would be far harder to accomplish than changing or dropping SIPP.

A second issue in the decision to drop SIPP was concern about its quality and usability. At the time of the proposed cut and even before, many SIPP users believed that the survey had developed serious problems, completely apart from the funding crisis (although some may have been related to previous economies in data collection and processing). The problems included sample attrition (explicitly cited by the Census Bureau in its announcement about dropping SIPP),[7] underreporting of both income and participation in various government programs, a lengthy lag between data collection and the availability of public-use files because of an outdated and cumbersome data editing and processing system, and the difficulty many users had in working with SIPP data files because of their complex structure and inadequate documentation (see "Strengths and Weaknesses of SIPP Data" below).

All of these problems had long been recognized and indeed were examined in detail in the 1993 report *The Future of the Survey of Income and Program Participation*. Although the Census Bureau had continued its attempts to address the problems and to act on the recommendations provided by the CNSTAT report, these efforts were at best only partially successful, and many of the earlier recommendations were essentially ignored, in large part because funding was not available to carry them out. Thus, with or without a funding crisis, the Census Bureau needed to perform nontrivial surgery on SIPP, and it was working on redesigning the data processing system to implement when a new panel began following the 2004 panel.

[7]E-mail memorandum from Carole Popoff to the Census Bureau's electronic mailing list for SIPP users, February 6, 2006.

The bureau's proposal for fiscal 2007 retained only $9.2 million of SIPP's full funding of about $44 million.[8] Of this amount, $3.6 million was to support continued data collection for the 2004 panel, although the Census Bureau said that either it would need additional funding from other agencies or else the 2004 panel would have to be terminated in September 2006 (the original plan was to continue it through the end of 2007).

The remaining $5.6 million in the bureau's proposed budget for fiscal 2007 would be used to design a new program to collect longitudinal information, dubbed shortly thereafter the Dynamics of Economic Well-being System (DEWS). The new system would rely much more heavily on administrative data for information on program participation and would markedly scale back the survey component. One thought was that an existing survey, such as the CPS, might provide baseline data for a sample cohort: current and retrospective income and program data would be obtained from administrative records, and additional information would be obtained from follow-up interviews at annual intervals. The original charge to our panel was to evaluate the plans to use administrative records for this new program.

When the Census Bureau announced in early 2006 that SIPP would be terminated and replaced with DEWS, the user community, led by researchers at the Center for Economic and Policy Research, reacted with unexpected speed and forcefulness. SIPP advocates sent letters to Congress arguing that SIPP was crucial to policy research and should be continued and that the Bush administration's recommended cuts in SIPP should not be implemented. One public letter, signed by over 400 researchers, including two Nobel laureates, was sent to Congress on March 1, 2006. The letter attested that SIPP "provides a constant stream of in-depth data that enables government, academic, and independent researchers to evaluate the effectiveness and improve the efficiency of several hundred billion dollars in spending on social programs" and that cutting the survey would lose the investment made over the years in collecting and using the data for important policy analysis and applied social science research.[9] The letters were accompanied by effective lobbying of Congress, especially of staff and members on the appropriations committees in the House and Senate that control the Census Bureau's budget. The lobbying campaign was assisted by a surprising level of coverage in the media, including an editorial in the *New York Times* on March 4, 2006, recommending that SIPP be retained.

---

[8]E-mail memorandum from Carole Popoff to the Census Bureau's electronic mailing list for SIPP users, February 6, 2006. SIPP had been receiving about $34 million in annual appropriations plus another $10 million originally appropriated for the completed SPD, which was reallocated to SIPP.

[9]Available at http://www.ceprdata.org/savesipp/resletter-name.pdf; see also Glenn (2006).

Reengineering (2006-Present)

Congress passed a fiscal 2007 budget in February 2006, in which it refused to accept the administration's proposal to terminate SIPP, although it did not restore full funding. Instead, Congress cut SIPP funding by about 25 percent, from about $44 million to $32.6 million (including $10 million from the appropriation that originally provided for the SPD). The Census Bureau in turn cut the 2004 SIPP panel sample size by over 50 percent, from 45,700 original sample households still eligible for the survey (this number includes new households formed by panel members after Wave 1) to 21,300 original sample households for the last four waves of the panel; it also eliminated the topical modules for these waves. This reduction allowed the agency to reduce SIPP spending of about $44 million annually to $25.4 million and to use part of the savings to continue disseminating data to users from earlier waves of the 2004 panel. The Census Bureau planned to use the remaining $7 million of the 2007 appropriation to work on developing the new DEWS program to replace SIPP. In effect both the advocates who wanted SIPP to continue and the Census Bureau and a portion of the user community who wanted to redesign SIPP got part of what they wanted: SIPP was continued, albeit with a reduced sample, and the Census Bureau continued work on developing the DEWS program.

With the restoration of funds for SIPP in the 2007 budget and again in the 2008 budget, the Census Bureau in September 2008 began a new panel under the existing design and processing system with a sample of about 45,000 households. In addition, at the instigation of Congress and data users, the bureau abandoned the DEWS concept of using administrative records in place of most survey content and instead embarked on a redesign or reengineering of SIPP. Thus, the report of the House Appropriations Committee on the Commerce, Justice, Science, and Related Agencies Appropriations Bill, 2008, issued July 19, 2007, directed "the Bureau of the Census to suspend activity on the DEWS survey development" and, instead, "to work with stakeholders to reengineer the SIPP to develop a more accurate and timely survey to capture the economic dynamics of the country."

The currently available funding is sufficient to continue the 2008 panel with a full sample. This level of funding also allows for work to go forward on reengineering the current SIPP, including the following components:

- improvements in the data collection instrument and processing system to achieve greater efficiency of operations and timeliness of data products, such as converting the current DOS-based software that supports computer-assisted interviewing for SIPP to a Windows-based system called BLAISE;

- development and evaluation of an event history calendar to facilitate collection of monthly core data in annual interviews;
- evaluation of administrative records data to supplement and evaluate the survey data; and
- development of survey content and use of reimbursable supplements, through interactions with stakeholders.

The goal is to implement the first 3- or 4-year panel under the new design in 2013. If the testing program supports it, the new design for SIPP panels will consist of three (or four) annual interview waves, each of which will collect data for the previous calendar year (using an event history calendar), with content similar to that collected in the current SIPP core questionnaire, plus some previously topical module content moved into the core. There will be no topical modules as such, but agencies can obtain additional information by paying for supplemental questions, which are most likely to be asked between the core interviews.

STRENGTHS AND WEAKNESSES OF SIPP DATA

Ideally, a reengineered SIPP would preserve or even enhance the survey's strengths while ameliorating many of its weaknesses. SIPP's principal strengths include

- its unique and extensive monthly data on employment, earnings, program participation, and household composition;
- the information collected on assets, shelter costs, medical expenses, and other items in its periodic topical modules that is necessary to simulate program eligibility and take-up rates;
- the detailed information collected on an array of subject areas related to socioeconomic well-being in its periodic topical modules; and
- the overall quality of the information collected on program participants and the low-income population generally relative to other household surveys.

SIPP's major weaknesses include

- a marked decline in the quality of income data as income rises;
- misplaced and erroneous transitions in income receipt, program participation, and health insurance coverage;
- possible biases arising from attrition and an underrepresentation of new entrants to the population (such as births, immigrants from abroad, and people moving from group quarters to household residences);

- a lack of timeliness in the release of data files; and
- until the late 1990s when the first edition of the *SIPP Users' Guide* was published, inadequate documentation to assist users in working with the complex SIPP public-use microdata files.[10]

SIPP panels are also shorter in length than panels in most other longitudinal data sets, which limits the usefulness of the information from SIPP for modeling long-run dynamics.

SIPP's Unique Value

SIPP stands alone among nationally representative household surveys in collecting income and program participation by month on a recurrent basis, and it does so at the person level for an extensive array of sources. Because of this feature, SIPP is uniquely able to support monthly estimates of participation in and eligibility for many federal and even state programs, although eligibility simulations still require imputation of components (such as assets, shelter costs, child care expenses, and other employment-related expenses) that are either not collected in the SIPP or are collected at times other than the month being estimated. SIPP is also unique in its ability to support models of short-term dynamics over a wide range of characteristics, including models of earnings dynamics based on its monthly data on employers and wages. The household component of the continuous MEPS—see http://www.meps.ahrq.gov/mepsweb/—also collects data on short-term dynamics of employment and health insurance coverage, in 5 interviews over a 2.5-year period for each panel, providing 2 calendar years of data. However, income data are collected only twice in MEPS panels, using a calendar-year reference period, and MEPS has a markedly smaller sample size than SIPP, even when two overlapping panels are combined for calendar-year estimates. MEPS also covers a shorter span of time than SIPP (2 years versus 3 or 4 years), which limits analysis of transitions that are experienced by only a small proportion of the population in a given year. SIPP's topical modules expand the survey's content to include types of data that few other surveys collect—such as wealth, child care and housing expenditures, and marital and immigration histories. SIPP's topical module data on disability have become the model of excellence for disability measurement.

---

[10]The first edition of the *SIPP Users' Guide* covered the 1984-1993 panels; it was updated through the 1996 panel in 2001 and is currently partially updated through the 2008 panel (see http://www.census.gov/sipp/usrguid.html).

Overall Quality of SIPP Income Data[11]

Assessments of data quality in a national survey such as SIPP typically rely on comparisons with other surveys or, for certain types of data, administrative records. Unless a particular other survey has been established as the gold standard in a given area—as is true, for example, of the Survey of Consumer Finances for the measurement of wealth—comparisons across surveys may indicate only where surveys differ and not which is best. Compounding the difficulty of evaluating SIPP data is the general uniqueness of SIPP's monthly estimates among surveys. The survey's great strength lies in collecting data that are not obtained elsewhere, but this limits how fully SIPP data can be evaluated.

No survey matches program administrative totals with respect to total recipients or, especially, aggregate dollars, but among the major national surveys SIPP performs best overall. For programs with high turnover, such as Medicaid, SIPP finds as many participants in a typical month as the CPS finds over a calendar year (Czajka and Denmead, 2008). This suggests that SIPP's superiority may be a direct result of its frequent interviews and short reference period, underscoring the challenge that the Census Bureau faces in planning to reduce three interviews per year to just one, with a 12-month reference period.

Compared with the CPS, the official source of income and poverty statistics for the United States, SIPP captures more income from families in the bottom quintile of the family income distribution, finds more sources of income and less reliance on Social Security among the elderly, and finds a somewhat smaller proportion of the population in poverty. Except for self-employment and entitlement programs, however, SIPP's superiority in the measurement of income is restricted to the bottom quintile. Overall, SIPP captures only 89 percent of the aggregate income recorded in the CPS, which in turn underestimates total household income in comparison to administrative records. The American Community Survey (ACS), which uses a mailout/mailback questionnaire to collect data from about half of its respondents, obtains 98 percent as much total income as the CPS (Czajka and Denmead, 2008).

Data Quality Shortcomings

With the monthly data collected in SIPP, users can estimate transitions involving a wide range of phenomena, including labor force activity, program participation, health insurance coverage, and family composition.

[11]An extended discussion of SIPP data quality, including additional citations, appears in Appendix A.

Estimates of the timing of transitions and the duration of spells created by transitions are affected by various types of reporting error that can generate a pronounced seam bias—that is, a tendency for transitions to fall disproportionately at the seams between waves rather than within the surrounding reference periods. In SIPP, transitions can occur between months 1 and 2, 2 and 3, or 3 and 4 of a 4-month reference period or between month 4 of one reference period and month 1 of the next reference period. SIPP's rotation group structure distributes interviews uniformly by calendar month, so changes in such characteristics as program participation, employment, and health insurance coverage should occur with the same frequency between any consecutive pair of reference months within or between survey waves. Instead, such transitions are more likely to be reported between month 4 of one wave and month 1 of the next wave than between any pair of months within the same wave. The extent of seam bias varies widely across characteristics but is particularly strong for health insurance coverage and program participation in general. For example, in one recent analysis of the 2001 SIPP panel, between 83 and 100 percent of transitions into or out of the major sources of health insurance coverage were reported at the seam between interviews (Czajka and Mabli, 2009).

While the likely causes of seam bias in panel surveys are many and varied (Callegaro, 2008), the principal source of seam bias in reported health insurance coverage in SIPP appears to be a tendency for respondents to report that they or other household members were covered by a particular source for either all 4 months or no months of the reference period. This phenomenon has a pronounced impact on distributions of duration. Excluding persons who were uninsured for all 36 months, 64 percent of the nonelderly adults who were uninsured for some portion of the 2001 panel were reported as uninsured for a multiple of 4 months (Czajka and Mabli, 2009).

While seam bias may pose a serious problem for longitudinal analysis with SIPP, its impact on cross-sectional estimates is muted by SIPP's rotation group design, which ensures that seams are distributed uniformly across calendar months. Monthly estimates will reflect any net reporting bias, but the bias for any survey wave will be distributed uniformly across the calendar months of the reference period. This is important for estimates of monthly program eligibility and participation.

Too Many Transitions?

Health insurance coverage estimates from SIPP illustrate the general problem of overstated transitions and their implications for longitudinal analysis. Average monthly estimates of health insurance coverage from SIPP compare closely with estimates of health insurance coverage obtained in the National Health Interview Survey, which measures coverage at the time of the interview (Czajka and Denmead, 2008; Davern et al., 2007). However, changes in coverage in SIPP occur with a frequency that strains belief—particularly among children. Among both adults and children, persons who experience changes in coverage often revert back to their original coverage at the start of the next wave, suggesting that reporting error may play an important role (Czajka and Mabli, 2009).

Attrition

Attrition is the bane of panel surveys, as more and more cases drop out because they move and cannot be found or refuse to stay in the survey. While SIPP enjoyed initial response rates at Wave 1 above 90 percent prior to the 2001 panel, the Wave 1 response rate dropped to 87 percent in the 2001 panel and 85 percent in the 2004 panel (see Table 2-1). Moreover, cumulative attrition has always been appreciable. In the 1996 panel, by the end of Wave 12, the cumulative sample loss—including the 8.4 percent initial Wave 1 nonresponse—exceeded 35 percent. With the discontinuation of a practice of terminating households that missed two consecutive interviews after Wave 1, the Census Bureau reduced the cumulative attrition rate at Wave 9 by 1 percentage point between the 1996 and 2001 panels. Nevertheless, cumulative attrition remains high (Czajka, Mabli, and Cody, 2008), and indeed increased for the 2004 panel. It should be noted that attrition is increasing over time with all household surveys.

Even more than its impact on sample size, attrition raises concerns because of its potential biasing effect. There is ample evidence from comparisons of characteristics measured in the initial waves of panel surveys that attriters differ from stayers. However, evidence using matched administrative records, which are not subject to differential reporting error between attriters and stayers, indicates that differences between the two groups diminish over time (Vaughan and Scheueren, 2002). In a long panel, even with no adjustment for differential attrition in the survey weights, cross-sectional bias will be reduced by this phenomenon, but the amount of change over time will be underestimated. Another study using the same sources of administrative records found that there were negligible differences between the stayers—reweighted to represent the Wave 1 universe—and the full, initial sample on annual earnings reported to the Internal Revenue Service (IRS), Social Security income and type of recipiency, and benefit amounts from the Supplemental Security Income (SSI) program (Czajka, Mabli, and Cody, 2008). Even estimates of change over time showed little evidence of bias. While limited to a small set of variables, these findings suggest that when respondents leaving the survey universe are handled appropriately and the Census Bureau's weighting adjustments are taken into account, the evidence of attrition bias in the SIPP is not as strong as is commonly assumed. Nevertheless, as long as attrition remains high, there is always reason to be concerned that the remaining sample cases may over or underrepresent particular types of people, events, or temporal phenomena—especially those associated with disruptions in personal circumstances.

Other Bias Concerns

Although SIPP is a panel survey, cross-sectional uses may be more common than longitudinal analyses of SIPP data. Evidence that cross-sectional estimates of poverty show trends that deviate from trends recorded in the CPS suggests a panel bias that should caution users against reliance on cross-sectional estimates from later waves (Czajka, Mabli, and Cody, 2008). If attrition is not the principal cause, then renewed efforts to understand the sources of the problem would benefit the survey redesign. A possible contributor to the problem of panel bias in cross-sectional estimates is SIPP's underrepresentation of persons who join the population after the initial interview (Czajka and Mabli, 2009).

Recent panel estimates show an appreciable reduction in poverty between Waves 1 and 2, yet little change over the next waves. Seeking an explanation in the first two waves, Czajka, Mabli, and Cody (2008) compared poverty status between the first two waves of the 2004 panel and found that changes in recorded poverty among persons present in both waves, rather than excess attrition among the Wave 1 poor, accounted for 87 percent of the net reduction in the number of poor. Did the experience of the Wave 1 interview make the respondents better reporters of income in Wave 2 (an example of time-in-sample bias), or is this nothing more than a classic regression to the mean? Whatever the cause or causes, the possibility that Wave 1 data behave differently from subsequent waves becomes a matter of greater concern if Wave 1 becomes the first of only 3 or 4 annual interviews rather than 1 of 12 part-year interviews.

Lack of Timeliness

One commonly articulated problem with SIPP data is the lag between when the data are collected and when they are released. For example, Wave 1 interviews of the 2004 SIPP panel were conducted between February and May 2004. Data collected in the core instrument were not released until late April 2006, an interval of nearly 2 years, with a re-release of the data to correct minor errors a few months later. Wave 2 core data were not released until March 2007, or 30 months after the interviews were completed. Even by Wave 6 of the 2004 panel, the lag between collection and release remained well over 2 years. Certainly, it should be recognized that the 2004 panel incorporated several changes over the previous (2001) panel that contributed to these delays. Nevertheless, the delays associated with the 2004 panel have been the norm for SIPP panels more often than the exception.

CONCLUSIONS AND RECOMMENDATIONS

Conclusion 2-1: The Survey of Income and Program Participation is a unique source of information for a representative sample of household members on the intrayear dynamics of income, employment, and program eligibility and participation, together with related demographic and socioeconomic characteristics. This information remains as vital today for evaluating and improving government programs addressed to social and economic needs of the U.S. population as it did when the survey began 25 years ago.

Conclusion 2-2: The Survey of Income and Program Participation's (SIPP) history of forward movement followed by setbacks has contributed to the survey's falling short of its original promise with regard to timeliness, usability, and maintenance of data quality. With the Census Bureau's planned SIPP reengineering program, there is an opportunity to put the survey on a much firmer foundation for the future. It is essential that the Census Bureau's program to reengineer SIPP address its problems and retain and build on its unique value and strengths.

No survey can be all things to all users. In reengineering SIPP, the focus should be on improving the content and design features of the survey that make possible its unique contribution.

Recommendation 2-1: To guide the design of a reengineered Survey of Income and Program Participation, the Census Bureau should consider the primary goal of the survey to be to provide data for policy analysis and research on the short-run (intrayear) dynamics of economic well-being for families and households, including employment, earnings, other income, and program eligibility and participation.

Recommendation 2-2: The Census Bureau's reengineering program for the Survey of Income and Program Participation should explicitly evaluate each proposed innovative feature, such as the use of administrative records or an event history calendar, on the extent to which a feature contributes to the survey's ability to measure short-term changes in economic well-being with improved quality and timeliness.

3

Expanded Use of Administrative Records

In reengineering the Survey of Income and Program Participation (SIPP), the Census Bureau has from the outset envisioned a role for administrative records. Although the bureau backed away from the notion of using administrative records to replace a large portion of the SIPP questionnaire content (see Chapter 2), it has continued to stress the contribution that administrative records could make to improving the quality of SIPP data (see Johnson, 2008).

This chapter addresses the role that administrative records can play in a reengineered SIPP. The chapter first outlines a framework for evaluating the benefits and costs of different uses of administrative records for SIPP. Using the framework as a guide, the chapter reviews the uses of administrative records in SIPP's history to date, along with other uses of administrative records at the Census Bureau that are relevant to SIPP. It then addresses the feasibility of acquiring and linking different federal and state administrative records and the benefits and costs of the following seven ways of using such records in a reengineered SIPP:

1. evaluating the accuracy of survey responses in the aggregate by comparison with aggregate estimates from administrative records;
2. evaluating the accuracy of survey responses at the individual respondent level by comparison with exactly matched administrative records;
3. improving the accuracy of imputation routines used to supply values for missing survey responses and of survey weighting factors used to improve coverage of the population;

4. providing values directly for missing survey responses;
5. adjusting survey responses for underreporting or overreporting;
6. using administrative records values instead of asking survey questions; and
7. appending administrative records values to survey records.

The first three uses we term "indirect," in that administrative data are never actually recorded on SIPP data files; the last four uses are "direct," in that administrative data become part of the SIPP data files to a greater or lesser extent.

Following the discussion of uses, the chapter considers methods of confidentiality protection and data access that would be appropriate for a reengineered SIPP. Our conclusions and recommendations are presented at the end of the chapter.

A FRAMEWORK FOR ASSESSING USES OF ADMINISTRATIVE RECORDS

SIPP's primary goal—which is to provide detailed information on the short-term dynamics of economic well-being for families and households, including employment, earnings, other income, and program eligibility and participation—requires a survey as the main source of data. There are no administrative records from federal or state agencies that, singly or in combination, could eliminate the need for survey data collection, even if it were feasible to obtain all relevant records and the custodial agencies did not object to their use for this purpose.

Consider the following examples of shortcomings in administrative records:

- Records for programs to assist low-income people, such as the Supplemental Security Income (SSI) Program or the Food Stamp Program (since 2008 termed the Supplemental Nutrition Assistance Program or SNAP), contain information only for beneficiaries and not also for people who are eligible for the program but do not apply for or are erroneously denied benefits. Being able to estimate the size of the eligible population, including participants and nonparticipants, is important to address the extent to which an eligible population's needs are being met, what kinds of people are more or less likely to participate in a program, and other policy-relevant questions.
- Program records do not always accurately distinguish new recipients of benefits from people who received benefits previously, had a spell of nonparticipation, and are once more receiving benefits. One of

SIPP's important contributions to welfare program policy analysis has been to make possible the identification of patterns of program participation over time, including single and multiple spells.

- Federal income tax records on earnings and other income exclude some important income sources that recipients do not have to report, such as Temporary Assistance for Needy Families (TANF) and pretax exclusions from gross wage and salary income. Pretax employer-sponsored health insurance contributions, which are a growing share of wage and salary income, do not have to be reported on Internal Revenue Service (IRS) 1040 individual income tax returns, nor are they always reported on W-2 wage and tax statements.
- Federal income tax records do not define some income sources in the manner that is most useful for assistance program policy analysis. Thus, self-employment income is reported to tax authorities as gross income minus expenses, including depreciation of buildings and equipment, which can result in a net loss, even when the business provided sufficient income to the owner(s) for living expenses. In contrast, the SIPP questionnaire asks for the "draw" that self-employed people take out of their business for their personal living expenses.
- The recipient or filing unit that is identified in administrative records often differs from the family or household unit that is of interest for policy analysis. For example, minor children may be claimed as dependents on the income tax return of the noncustodial parent, and unmarried cohabitors will be two distinct income tax filing units but only one survey household and (assuming they share cooking facilities) one food stamp household. (It is not always possible to accurately identify tax and transfer program filing units in survey data, either.)

Despite these and other problems, it is clearly the case, as we demonstrate in later sections, that administrative records can be helpful to SIPP in a number of ways, as they have been helpful in the past (see "SIPP's History with Administrative Records" below). Indeed, the Census Bureau hopes that significantly greater use of administrative records can be achieved in a reengineered SIPP to improve the quality of reporting of income and program participation.

The benefits and costs of using administrative records for a reengineered SIPP must be carefully assessed, and each of the possible seven uses identified above implies a different mixture of benefits and costs. We provide below a cost-benefit framework for considering alternative uses of administrative records for SIPP, including not only records from federal agencies, but also records that state agencies use to administer such programs as the Children's Health Insurance Program (CHIP), food stamps, general assistance,[1] Medicaid, school lunch and breakfast programs, the Special Supplemental Nutrition Program for Women, Infants, and Children (WIC), SSI (in states that supplement federal benefits), TANF, unemployment insurance (UI), and workers' compensation (WC) (referred to as "state records" in this chapter).

Benefits

There are potentially two types of benefits for a reengineered SIPP from using administrative records, such as Social Security payments to beneficiaries or food stamp allotments to families: (1) providing higher-quality data in comparison to survey reports (the one benefit specifically identified by the Census Bureau) and (2) providing additional data that would be more difficult or expensive to obtain in interviews. For improving data quality, administrative records may also have the advantage that the ongoing costs of using them for this purpose are modest—at least once an initial investment has been made in acquiring and processing them—compared with efforts to improve the quality of survey reporting (see "Costs" below).

Improved Data Quality

There is substantial evidence, summarized in Chapter 2, that survey reports of program participation and sources of income are often incomplete and inaccurate—despite considerable efforts to improve the quality of reporting by redesigning questions, adding probes, and the like. SIPP, with its detailed, probing questionnaire, historically has a record of obtaining more complete reporting of program participation than other surveys, but its reporting of program participation still falls short of administrative benchmarks. Moreover, the amounts reported by acknowledged participants often differ from administrative benchmarks in the aggregate and on an individual basis. There are both underreporting and overreporting errors, typically with a net underreporting on balance. Consequently, administrative records have the potential to provide significantly more accurate data on many sources of income and types of programs.

In assessing the benefits of improved data quality from using a particular administrative records source, such as Social Security or food stamp records, it is important not to take at face value that the administrative record is always of better quality than the corresponding survey response.

[1]General assistance, or general relief, is a name for state programs to provide cash benefits to adults without dependent children.

In this regard, it is important to distinguish among the data items recorded on administrative records. On one hand, for example, in the case of a record for a food stamp recipient, it is highly likely that the amount provided to a beneficiary is accurately recorded (even though, in some cases, the payment may have been made to someone who was not in fact eligible for the program or an erroneous amount may have been provided to an eligible recipient). On the other hand, the ancillary information on the record, such as the person's employment, income, and family composition, may have contained errors when it was collected or may have become out of date. Moreover, for some programs, records for people who no longer receive benefits may be comingled with records for current beneficiaries, and, for most if not all programs, the program unit of one, two, or more people is typically not the same as the survey unit of a family or household.

The information in administrative records, even when accurate, may differ sufficiently in definition from the information sought by the survey designer as to make the administrative information unusable for the survey's purpose. As noted earlier, self-employment income from federal income tax records is an example—although the gross and net income amounts from tax records may be of interest for some analyses, they do not satisfy SIPP's purposes of understanding the economic resources available to individuals and families.

Additional Data

Some administrative records may contain valuable information that would be difficult to obtain in a survey context. For example, the Social Security Administration (SSA) has records not only of benefits paid to retirees, people with disabilities, and others, but also histories of earnings received each year for everyone who is or has been in covered employment, which SSA receives annually from W-2 and Self-Employment Income forms filed with the IRS.[2] Such earnings histories, which may extend back for decades of an individual's work life, would be difficult to collect in a survey unless it began following individuals from an early age, but they could be valuable for some types of research, such as research on the determinants of the decision to retire.

Costs

The use of administrative records for a reengineered SIPP cannot be cost free. Staff time and other resources must be expended for acquisition

[2]Prior to 1978, SSA files contain quarterly indicators of covered employment in addition to annual earnings.

and processing of records. Moreover, the use of some kinds of records could potentially incur two other types of costs: (1) increased delays in releasing data products due to delays in obtaining records from the cognizant agencies and (2) increased risks of disclosure of individuals in SIPP, which in turn could necessitate more restricted conditions for use of the data.

Additional Resources

The strictly monetary costs of using administrative records for a reengineered SIPP would include staff and other resources for acquisition of records, data quality review and associated cleaning of records, and processing of records for the particular application, such as evaluation or imputation. In some cases, the costs of acquisition could be substantial, at least initially. For example, time-consuming negotiations could be required to draw up acceptable memoranda of understanding and other legal documents to obtain an agency's records, although once agreed-upon procedures were in place, the marginal costs of acquiring records in subsequent years could be minimal. There could also be significant costs when an agency's records are not well maintained, requiring Census Bureau staff to engage in substantial back-and-forth with agency staff to clean up the data. Processing costs would vary with the type of application. For example, aggregate comparisons of survey responses with administrative records are likely to be considerably less costly than the use of administrative records in imputation models.

In its original concept for a new Dynamics of Well-being System (DEWS), the Census Bureau had hoped that administrative records could be used directly to supply so much of the needed content as to make possible a significant reduction in the costs of the system compared with the current SIPP. The cost savings would come from reduced frequency of interviews and reduced content of each interview, with the remaining needed content obtained by matching administrative records for individuals to the corresponding survey records. However, users were concerned that such a major role for administrative records would not only be unfeasible, given the difficulties of acquiring all of the needed records from state and federal agencies, but also would curtail the bureau's ability to release public-use microdata files because of increased disclosure risk. These concerns led the bureau to scale back its plans in this regard. The Census Bureau now plans to achieve cost savings by conducting annual interviews with event history calendars to obtain intrayear information and by requiring agencies to pay for supplements with variables not included in the core questionnaire (see Chapter 4).

Reducing the frequency of interviews assuredly reduces the costs of a survey, but whether reducing the content of a particular interview by substituting administrative records reduces costs is not clear. The main cost of an interview is making contact with the respondent; moreover, acquiring and processing administrative records adds costs. Hence, we think that the use of administrative records to replace survey content should be judged primarily on criteria other than cost savings, such as the effects on data quality, timeliness, and accessibility.

Increased Delays

Administrative records systems are managed first and foremost to facilitate the operation of assistance programs. The Census Bureau's need for timely information from records systems for statistical purposes is of secondary importance, at best, for program agencies. Consequently, while it may be possible for the Census Bureau to obtain and process some records with little delay, the acquisition of other records may lag the survey data collection by significant periods of time (see "Statistical Administrative Records System" below). One response to this situation could be to further delay the data products from SIPP in order to be able to use the administrative information to improve imputations or substitute for questionnaire content. This outcome would be distressing to users. Other responses could be to project the administrative information from a prior year forward to the survey data year, to issue preliminary and revised data products, or to confine the use of administrative information to evaluation of the survey content, which would not be as time sensitive.

Increased Disclosure Risks

On one hand, because the data collected in SIPP is of great interest to policy analysts, researchers, and others users, it is essential to make the data in some form available to these varied constituencies. On the other hand, the Census Bureau is ethically and legally obligated to protect the confidentiality of SIPP participants' identities and attributes. Thus, unfettered access to all collected SIPP data is not likely to be achievable. Rather, as recommended in previous National Research Council panels on data access (2005, 2007), an appropriate strategy for the Census Bureau is to provide access to data of differential detail, and hence differential disclosure risk, depending on the goals for data use and the trustworthiness of the likely data users (see Box 3-1 for a summary of the risk and utility trade-off in data dissemination).[3]

[3]We do not discuss confidentiality threats that might originate from inside the Census Bureau. The bureau has sufficient expertise on internal confidentiality protection that it does not need our panel to comment. Evidence of its dedication to confidentiality protection is the practice adopted for its Statistical Administrative Records System of substituting personal identification keys for Social Security numbers on matched files.

BOX 3-1
The Risk and Utility Trade-Off in Data Dissemination

The Census Bureau and other disseminators of data collected under a pledge of confidentiality for statistical purposes strive to release data that are (1) safe from attacks by ill-intentioned data users seeking to learn respondents' identities or sensitive attributes, (2) informative for a wide range of statistical analyses, and (3) easy for users to analyze with standard statistical methods (Reiter, 2004). These goals are often in conflict. For example, releasing fine details about individuals enables accurate analyses, but it also provides ill-intentioned users with more and higher quality resources for linking records in released data sets to records in other databases. Releasing highly aggregated summaries of data protects confidentiality, but it severely limits the analyses that can be done with the data. Data disseminators usually choose policies that lie in between these two extremes, sacrificing absolute protection (possible only when not releasing any data) and perfect data usefulness (possible only when releasing all data as collected) for a compromise.

Most data disseminators are concerned with two types of disclosures. One type is *identity disclosure*, which occurs when ill-intentioned users correctly identify individual records using the released data. Efforts to quantify identity disclosure risk in microdata (records for individual respondents) generally fall into two broad categories: (1) estimating the number of records in the released data that are unique records in the population and (2) estimating the probabilities that users of the released data could determine the identities of the records in the released data by using the information in that data. The other type is *attribute disclosure*, which occurs when ill-intentioned users learn the values of sensitive variables for individual records in the data set. Quantification of attribute disclosure risk is often folded into the quantification of identity disclosure risk, since ill-intentioned users

SIPP's great value for policy analysis and research on short-term dynamics of economic well-being requires that users have access to microdata and not only aggregate summaries. Administrative records could potentially add valuable information to SIPP microdata, but the more information that is added, the greater the risk that individuals in the SIPP sample could be identified in public-use microdata files. Disclosure risk is also increased because people in the agency supplying the administrative data have knowledge that could be used to identify individuals in SIPP files. Countering such increased risk could require the use of disclosure protection techniques that would diminish the value of the public-use microdata products and compel users who require the confidential data for their research to seek access to one of the Census Bureau's Research Data Centers (RDCs). Yet for policy analysis that is in any way time sensitive, typically need to identify individuals before learning their attributes. Other types of disclosures include perceived identification disclosure, which occurs when intruders incorrectly identify individual records in the database, and inferential disclosure, which occurs when intruders accurately predict sensitive attributes in the data set using the released data. For a discussion of metrics for quantifying identification and attribute disclosure risks, see Duncan and Lambert (1989), Federal Committee on Statistical Methodology (1994), Lambert (1993), National Research Council (2005, 2007), and Reiter (2005).

Agencies must also consider the usefulness of the released data, often called data utility. Existing utility measures are of two types: (1) comparisons of broad differences between the original and released data and (2) comparisons of specific estimates computed with the original and released data. Broad difference measures are based on statistical distances between the original and released data, for example, differences in distributions of variables. Comparison of specific models is often done informally. For example, data disseminators look at the similarity of point estimates and standard errors of regression coefficients after fitting the same regression on the original data and on the data proposed for release.

Ideally, the agency releasing data optimizes the trade-off between disclosure risk and data utility when selecting a dissemination strategy. To do so, the agency can make a scatter plot of the quantified measures of disclosure risk and data utility for candidate releases. This has been termed the "R-U confidentiality map" in the statistical literature (Duncan, Keller-McNulty, and Stokes, 2001). Making this map can enable data disseminators to eliminate policies with risk-utility profiles that are dominated by other policies (e.g., between two policies with the same disclosure risk, select the one with higher data utility).

the alternative of accessing microdata in an RDC is daunting because it adds delays in making a successful application to the delays that are already incurred in release of the files from the Census Bureau.

A related risk of directly using administrative data in SIPP could be a decline in the willingness of people to participate in the survey once they were made aware of the planned uses of their administrative records. However, when 2004 SIPP panel respondents were informed halfway through the panel that administrative records might be used to reduce the need to ask them so many questions, less than one-half of 1 percent requested that record matches not be made for them (David Johnson, chief, Housing and Household Economic Statistics Division, U.S. Census Bureau, personal communication to panel, February 3, 2009; see also "Direct Uses" below).

Trading Off Benefits and Costs

Different types of uses of administrative records in a reengineered SIPP will present different pictures of the likely benefits and costs. For a given use, the benefits and costs may also differ by the type of record or even by the agency responsible for the record. For example, program agencies in some states may be more willing to share records with the Census Bureau for use with SIPP than with agencies in other states.

In determining when a particular use of a specific type of record warrants the investment, it is important always to bear in mind the goals of SIPP and that it cannot be all things to all users. For example, while SSA records of past earnings histories would be useful for research on lifetime patterns of employment and related issues, they might not contribute greatly to SIPP's primary focus on the short-term dynamics of economic well-being. Moreover, the addition of earnings histories to SIPP would substantially increase the risks of disclosure and consequently the need to restrict the use of data products containing them (see "SIPP Gold Standard Project" below). Some of the trade-offs involved in working with different types of administrative records for different purposes become evident in reviewing the history of uses of administrative records in SIPP and other Census Bureau programs.

SIPP'S HISTORY WITH ADMINISTRATIVE RECORDS

In order to achieve SIPP's goals of improving information on the economic well-being of the population and short-term changes in income and program participation, the survey's designers at the outset envisioned at least three major roles for administrative records (see National Research Council, 1993:31-33):

1. to increase sampling efficiency by providing supplementary frames of participants in specific assistance programs or persons with other specified characteristics;
2. to provide additional data (e.g., by matching with Social Security earnings records to obtain longitudinal earnings histories to add to the SIPP files); and
3. to compare and validate specific items common to both SIPP and administrative records by means of record-check studies.

ISDP Use of Records

The Income Survey Development Program (ISDP) used administrative records extensively to evaluate the quality of survey responses and to improve question wording and interviewer training procedures (see Kasprzyk, 1983; Logan, Kasprzyk, and Cavanaugh, 1988). The primary method used was the forward record check, in which people included in independent samples from administrative sources (including IRS and federal and state program records) were administered the ISDP interviews. This method eliminates the need to match survey and administrative records, but it permits only identifying false-negative responses (people with an administrative record of program participation who say they did not participate in the particular program) and not also false-positive ones, which a full record-check study would support. Aggregate comparisons of income and program participation reported in the 1979 ISDP panel with administrative records sources were also conducted. These comparisons necessitated, in many cases, extensive adjustments of one or both sources (SIPP or the applicable administrative records source) for comparability of the population and income concept and reporting period covered.

The ISDP also drew supplementary samples from administrative records to augment the 1978 and 1979 ISDP panel main samples. However, the data were never analyzed, because data files that included the main and supplementary samples with appropriate weights could not be produced before the ISDP was shut down in 1981 (see Kasprzyk, 1983).

SIPP's Use of Records, 1983-1993

During SIPP's first decade, the Census Bureau was hard-pressed to operate the survey in full production mode and to accommodate budget reductions that necessitated cutbacks in sample size or number of interview waves or both for most panels (see Chapter 2). Bureau staff had limited time and resources to exploit the potential value of administrative records. Consequently, no supplementary sampling frames were developed from administrative records for SIPP during this period although some work went forward on evaluation and related uses of administrative records.

The Census Bureau carried out a handful of matches of SIPP panels with administrative records, which were facilitated by a successful program to obtain Social Security numbers from SIPP respondents and match them to SSA files for validation purposes. These matches included (1) a match of the 1984 SIPP panel with SSA records conducted for SSA under an agreement that limited its use to SSA analysts for a 2-year period; (2) a match of a small number of variables in IRS tax records with the 1984 panel conducted as part of an effort (which did not come to fruition) to develop weighting factors from IRS tax records for reducing the variance of income estimates from SIPP (Huggins and Fay, 1988); and (3) a match of IRS tax records with the 1990 panel conducted as part of an effort to develop a simulation model for estimating after-tax income in SIPP (which also did not come to fruition). An analysis of the 1990 panel-IRS match for married couples with earnings highlighted the contribution of imputation procedures to the long-standing pattern by which SIPP estimates of earnings have fallen short of IRS estimates in the aggregate (Coder, 1992). Another analysis used the 1990 matched file to estimate the extent to which eligible families applied for and received the earned income tax credit (Scholz, 1994).

Census Bureau staff also performed aggregate comparisons of selected estimates from administrative records sources and SIPP. Such comparisons were made for the 1984 panel for aggregate income amounts for nine sources (Jabine, King, and Petroni, 1990:Tables 10.1, 10.2, 10.3); for the 1986-1987 panels for the value of several types of assets and liabilities (Eargle, 1990: Table D.2); and for the 1990 panel for recipients and aggregate amounts for about 20 sources of income (Coder and Scoon-Rogers, 1996).

Bureau staff carried out a single full record-check study, which matched SIPP records in four states for the first two waves of the 1984 panel with records from eight federal and state programs—Aid to Families with Dependent Children (AFDC, the predecessor to TANF), food stamps, unemployment insurance, workers' compensation, federal civil service retirement, Social Security, SSI, and veterans' pensions and compensation. The study was designed to identify both false-negative and false-positive reports of program participation and benefit amounts in SIPP (Marquis and Moore, 1989, 1990a, 1990b). It encountered serious delays because of the time required to negotiate the acquisition of records from state agencies (indeed, the Census Bureau was never able to obtain the requested records from one state) and also because of problems in conducting the matches and preparing analysis files. Almost 5 years elapsed from the study's initiation in 1984 to the publication of detailed results, and many potentially useful analyses were never undertaken—in particular, the study did not examine discrepancies in benefit amounts but only in program participation.

Nonetheless, the SIPP record-check study made important contributions. Regarding the seam bias problem (see Chapter 2), it found that, in general, SIPP nonseam change estimates tended to underestimate true change, and change estimates at the seam tended to be too high. Regarding reporting bias, it confirmed the results of aggregate comparisons that participation in most programs was underreported (although there were overreports as well). It also found confusion among programs on the part of respondents, such as confusing AFDC with general assistance or Social Security with SSI. These findings stimulated research on questionnaire design to improve reporting in the survey.[4]

[4]Hotz and Scholz (2002:275-315), in a comprehensive review of surveys and administrative records sources for measuring employment and income of low-income populations, discuss some of the studies cited in the text.

SIPP's Use of Records, 1996-2006

Census Bureau staff have performed only one study comparing SIPP aggregates with independent estimates in the past decade. In this study, Roemer (2000) compared aggregate amounts for 1990-1996 from the 1990, 1991, 1992, 1993, and 1996 panels for about 16 sources of income with benchmarks derived from the National Income and Product Accounts and from the Current Population Survey (CPS). Recently, Meyer, Mok, and Sullivan (2009), researchers at the Universities of Chicago, Northwestern, and Notre Dame, compared aggregate amounts for nine income assistance programs for five surveys, including SIPP, the American Community Survey (ACS), the Consumer Expenditure Survey, the CPS, and the Panel Study of Income Dynamics (PSID), for years extending as far back as data were available (1983-2005 for SIPP). They also compared average monthly participation for eight of the nine programs for SIPP, the ACS, CPS, and PSID.

The SSA sponsored SIPP interviews in January 2003 and January 2005 for systematic samples drawn from its records of SSI recipients and Disability Insurance beneficiaries (supplementary sampling frames). The Census Bureau conducted the interviews with these beneficiaries and processed the interviews using standard SIPP procedures. These supplemental SIPP interviews are for use by SSA only and are not publicly available from the Census Bureau (see DeCesaro and Hemmeter, 2008, for a description of characteristics for SSI recipients and disability insurance beneficiaries from the SIPP January 2003 supplemental sample interviews).

The Census Bureau has continued to provide exact matches of SIPP and SSA records for use by SSA staff for research and modeling of the Old-Age, Survivors, and Disability Insurance (OASDI) and SSI Programs. The availability of these files not only has enabled SSA staff to conduct policy research that contributes to planning for OASDI and SSI Program needs (see, e.g., Butricia, Iams, and Smith, 2003; Iams and Sandell, 1996), but also has supported studies on the quality of SIPP reporting of OASDI and SSI benefits. Thus, Huynh, Rupp, and Sears (2001) used OASDI and SSI administrative records matched to the 1993 and 1996 SIPP panels to assess discrepancies in SIPP reports of benefit receipt and benefit amounts for four sample months. They found quite accurate reporting by people who received only Social Security and by people who received no benefits from either Social Security or SSI. However, there was substantial underreporting by people who received only SSI and by people who received both Social Security and SSI. They also found confusion between Social Security and SSI and much higher errors for imputed compared with reported benefit amounts for both programs.

Roemer (2002) used an exact match of the Detailed Earnings Record (DER) file from SSA with SIPP and CPS data for 1990, 1993, and 1996 to study the accuracy of annual estimates of wages from both surveys. He found net underreporting in SIPP along with reporting errors in both positive and negative directions: Fully 75 percent of SIPP respondents in the study reported wages that differed from the DER amount by more than 5 percent, although the correspondence was substantially better when comparing percentile ranks—only 37 percent of SIPP respondents differed in percentile rank between the two sources. Stinson (2008) developed a model of measurement error in both SIPP and the DER by analyzing matched cases from the 1996 SIPP panel.

Use of Records for Reengineering SIPP

The ongoing effort to reengineer SIPP at the Census Bureau is making use of two administrative records projects. They are the SIPP "gold standard" project and planned matches of administrative records from Illinois, Texas, and possibly other states with survey data.

SIPP Gold Standard Project

Begun in 2002 with funding from the Census Bureau, SSA, and the National Science Foundation, the goal of the SIPP Gold Standard project is to develop a rich resource for retirement income and disability analysis that can be widely used. The gold standard file, which can be analyzed only at the Census Bureau, includes variables from the 1990, 1991, 1992, 1993, and 1996 SIPP panels matched with IRS summary earnings records (annual FICA-taxable earnings, 1937-2003), IRS detailed earnings records (annual job-level data, uncapped, 1978-2003), and SSA benefits data through 2002 from the Master Beneficiary Record, Supplemental Security Income Record, and Payment History Update System 831 file (Abowd, 2007).[5]

A prototype public-use Version 4.1 of the gold standard file is available for researcher use as a beta test file through application to SSA, with the promise that the Census Bureau will run the researcher's application on the gold standard file for comparison purposes. Version 4.1 contains all person-level SIPP and IRS variables from the Gold Standard Version 4.0, plus the benefit and type of benefit for a person's initial SSA benefit (if any), as of April 1, 2000. There are 16 replicates of the 4.1 file, representing four different sets of imputations for missing data and four different syntheses of selected variables for each set of imputations to protect confidentiality. Each

[5]The gold standard project refers to earnings histories as IRS records because they are provided to IRS as well as to SSA by employers. The very stringent confidentiality provisions of Title 26 and related regulations apply to the earnings data whether they are obtained from SSA or from IRS.

replicate has a consistent panel weight for the civilian, noninstitutionalized population as of April 1, 2000.

The ultimate goal of this work is to create public-use files that not only are useful for research on retirement and disability, but also protect against identifying SIPP respondents in the already available public-use files by applying state-of-the-art synthesizing techniques to selected variables. Such techniques perturb or alter specified variables according to specified statistical models that are designed to preserve key univariate and multivariate distributions to the extent possible. The challenge for synthetic techniques is whether they can fully protect confidentiality and at the same time permit inferences from the data that are as valid as would be obtained from a gold standard file. Initial evaluations of Version 4.1, which incorporated synthesized values for the vast majority of variables, showed excellent results for estimates of earnings histories for white men and women, not quite-so-good results for estimates of earnings histories for black men and women, and underestimation of early retirement and also of retirement at age 65 compared with other years (Abowd, 2007).

SIPP reengineering staff are interested to learn about the experiences and reactions of researchers who work with Version 4.1 to determine if this approach would be acceptable to the SIPP user community as a way to provide SIPP public-use files that are enriched with administrative records data. To date, half a dozen researchers are working with the beta file, and SSA commissioned an in-depth evaluation of it, which was completed in spring 2009 (see Urban Institute/NORC Evaluation Team, 2009). The conclusions of this evaluation are discussed in "Adding New Variables" below.

Illinois and Texas Matching Project

Work with Illinois and Texas program records began prior to the SIPP reengineering effort with a project to match 1999-2003 subsidized child care and TANF files from the two states with Census Bureau survey data as part of a study funded by the U.S. Department of Health and Human Services. In June 2008, the Census Bureau entered into agreements with both states to obtain administrative records for a Demonstration of Administrative Records Improving Surveys (DARIS) project. The goals of DARIS (see University of Texas and U.S. Census Bureau, 2008:3) are to "demonstrate methods of integrating data from surveys and administrative records, produce data sets that more accurately represent the target population's characteristics than survey data alone, conduct experiments in disclosure-proofing hybrid data sets, and document feasibility." The files provided by Illinois and Texas include the previously provided child care and TANF files for 1999-2003, extended through 2007, and food stamp participation files for 2004-2007.

The SIPP reengineering effort is taking advantage of the DARIS project to evaluate the quality of the responses obtained for a sample of 2004 SIPP panel members in Illinois and Texas who were interviewed in spring 2008 using a paper and pencil event history calendar to obtain information for calendar year 2007. This evaluation sample includes SIPP panel members who were dropped from the survey for budgetary reasons as well as continuing panel members who provided responses covering 2007. The administrative records data for 2007 for the evaluation sample are being compared with the responses in regular SIPP interviews covering 2007 (for the continuing panel members) and with the responses obtained in event history calendar test interviews (for continuing and dropped panel members). In a subsequent evaluation of an electronic event history calendar test, scheduled for early 2010, the Census Bureau hopes to compare the survey results with administrative records data not only from Illinois and Texas, but also from other states including Maryland (with which the Census Bureau already has an arrangement for obtaining program records—see below) and California, Massachusetts, New York, and Wisconsin. We discuss the event history calendar tests in Chapter 4.

Although not directly related to SIPP, we note that two other projects have demonstrated the value of exact matches of state administrative records with survey responses for evaluation purposes. For one project, the Census Bureau exactly matched administrative records from Maryland's Client Automated Resource and Eligibility System (CARES), which contains records for beneficiaries of food stamps, TANF, and several other public assistance programs, with the 2001 test version of the American Community Survey. Analysis of the matched files documented significant underreporting of program participation in the 2001 ACS (Lynch et al., 2008; Taeuber et al., 2004). The work was sponsored by the Economic Research Service, U.S. Department of Agriculture, and the Maryland Department of Human Resources. For the other project, the Census Bureau exactly matched administrative records from Illinois, Maryland, and Texas, including TANF records, child care records, and employment and earnings records, with the 2001 test version of the ACS. The analysis examined child care subsidy participation and the effects on employment among low-income families in the three states (Goerge, 2009).[6] This work was funded by the Child Care

[6]Illinois, Maryland, and Texas, along with six other states (California, Florida, Georgia, Missouri, Ohio, and Washington) participate in the Administrative Data Research and Evaluation (ADARE) alliance. ADARE is a partnership among research organizations, which have developed data-sharing agreements with their respective states to obtain administrative records databases for the TANF, Unemployment Insurance, Workforce Investment Act, and other employment-related programs for employment- and welfare-related research and evaluation. ADARE is funded by the Employment and Training Administration, U.S. Department of Labor, and managed by the Jacob France Institute at the University of Baltimore (see http://www.ubalt.edu/jfi/adare/about-ADARE.cfm).

Bureau of the Administration for Children and Families, U.S. Department of Health and Human Services.

OTHER CENSUS BUREAU USES OF ADMINISTRATIVE RECORDS

The Census Bureau has increasingly made use of administrative records in other programs, and many of these uses are relevant to a reengineered SIPP. Three major programs are briefly described below: the Longitudinal Employer-Household Dynamics (LEHD) Program; the Small-Area Income and Poverty Estimates/Small-Area Health Insurance Estimates (SAIPE/SAHIE) Programs; and the Statistical Administrative Records System (StARS).

Longitudinal Employer-Household Dynamics

LEHD is a 10-year-old program, supported by the Census Bureau, the National Science Foundation, the National Institute on Aging, and the Sloan Foundation, which seeks to link the Census Bureau's household and business surveys in ways that can advance knowledge of the dynamic relationships of workers, jobs, households, and businesses. A component of LEHD is the Local Employment Dynamics Program, in which the Census Bureau obtains quarterly employment and earnings information from state employment security agencies and, in return, provides quarterly workforce indicators (QWI) for labor market areas in each state. The states collect employment and earnings from almost all employers in order to manage their unemployment insurance programs; the QWI data are developed by the Census Bureau by merging local demographic information with the employment and earnings information (see http://lehd.did.census.gov/led/led/led.html).

As of early 2009, 47 states (excluding only Connecticut, Massachusetts, and New Hampshire), the District of Columbia, and Puerto Rico are or are about to be part of the LEHD Program through separate memoranda of understanding between each state and the Census Bureau. Researchers have made extensive use of LEHD information linked across time and other LEHD data sets for innovative analyses that have enriched understanding of labor markets (see, e.g., Brown, Haltiwanger, and Lane, 2006). Not only does the LEHD Program provide information on employment and earnings that could potentially be used to evaluate or augment SIPP data, but also the history of the initiation and growth of the program from a handful of states to its present almost-complete coverage may hold lessons for a reengineered SIPP (see "Acquisition of Administrative Records, State Records" below).

Small-Area Income and Poverty Estimates/ Small-Area Health Insurance Estimates

The Census Bureau, with support from other federal agencies, created the Small Area Income and Poverty Estimates Program in the mid-1990s to provide more current estimates of selected income and poverty statistics (e.g., poor, school-age children) than those from the most recent decennial census for small geographic areas. The program creates estimates for school districts, counties, and states using statistical models that incorporate data from the ACS (beginning with the 2005 estimates; previously, CPS data were used), together with administrative records data on food stamp recipients and federal income tax filers at the county level. The estimates are used in allocation of federal education funds to local jurisdictions. More recently, the Census Bureau began the Small-Area Health Insurance Estimates Program to provide state and county estimates of health insurance coverage using similar statistical models with CPS data and administrative records data for counties on food stamp recipients, federal income tax filers, and enrollees in Medicaid and CHIP.

SAIPE and SAHIE are examples of estimates that do not rely on a single source—survey or administrative records—but instead combine data from multiple sources in statistical models to reduce sampling and nonsampling errors in the estimates (see http://www.census.gov/did/www/saipe/ and http://www.census.gov/did/www/sahie/; see also National Research Council, 2001). It is possible that statistical models could be used to develop "best estimates" of selected key indicators, such as poverty rates, from SIPP (or other surveys), but we do not discuss this approach further.

Statistical Administrative Records System

In the early 1990s the Census Bureau began a program to develop an integrated set of administrative records that could be used for a variety of purposes to reduce reporting burden and to minimize the cost of obtaining needed information. The bureau inventoried potentially available administrative records files and created an administrative records research staff. The staff built a prototype of a combined and unduplicated set of administrative records (StARS 1999) that would include basic demographic information (age, race, ethnicity, and gender) similar to the decennial census short-form content. One of the 2000 census experiments compared census counts with estimates of population and demographic characteristics for census tracts and blocks in five counties derived from the 1999 StARS (National Research Council, 2004b:199-202).

Following the 2000 census, Census Bureau staff developed a model for imputing race and ethnicity from 2000 census data to improve on the available information in the Census Numident (numeral identification) file, which in turn is used to input demographic information to StARS. The Census Numident file is an edited version of the SSA Numident file that stores information contained in applications for Social Security numbers (SSNs), including the name of the applicant, place and date of birth, and other information for all SSNs since the first number was issued in 1936.

In addition, Census Bureau staff built a person validation system (PVS) that can match and verify records containing SSNs against the Census Numident file, or, if the records do not contain SSNs, determine a valid SSN either by matching on address against the geokey reference file, or by matching on name and date of birth against the name reference file. The geokey reference file is generated from StARS and contains all addresses for each SSN; the name reference file is also generated from StARS and contains all combinations of alternate names and dates of birth for each SSN. The PVS replaces the SSNs with person identification keys (PIKs) to enhance the level of confidentiality protection.

The PVS system is very important for SIPP, which stopped collecting SSNs midway through the 2004 panel because of increasingly poor response. An evaluation of the PVS using CPS 2001 records—47 percent of which lacked SSNs—found that the PVS achieved a verified matching rate for the total CPS sample of 93 percent, using address, name, and date of birth, compared with a rate of 94 percent when SSNs were also used in the match when available (Wagner, 2007:slide 14—the match excluded CPS records with no name and refusals).

Once a set of records, such as SIPP survey responses, has been matched via the PVS, it is then possible to use the resulting PIKs to match the survey records with other records that the Census Bureau has acquired as part of its initiative to integrate and make better use of administrative records. The core StARS, which is designed to contain short-form-only content, at present includes over 300 million person records and over 150 million address records developed by merging and unduplicating seven national files. These files are IRS 1040 records, IRS 1099 records, and Medicare Part B records, along with two sets of records from the U.S. Department of Housing and Urban Development, a set of records from the Indian Health Service, and the Selective Service System registration file. In addition, the Census Bureau regularly acquires Master Beneficiary Record files from SSA for survey records to which it has assigned PIKs (and could acquire the complete files if so desired), Medicaid files from the Centers for Medicare and Medicaid Services, and SSI record files from SSA, along with quarterly wage records from states that participate in the LEHD Program and counts of food stamp recipients by county.

Applications of StARS and other administrative records acquired by the Census Bureau to date include

- research on using StARS records to assign age, race, gender, and Hispanic origin for census respondents who fail to report one or more of these characteristics;
- research on using StARS records to determine the demographic characteristics of households that do not respond to the CPS;
- work to develop near-real-time population estimates for areas that experienced disasters, such as a devastating hurricane—for this purpose, the Census Bureau acquired the U.S. Postal Service's National Change of Address File and the Federal Emergency Management Administration's emergency management and flood insurance files; and
- work to match Medicare and Medicaid files to CPS ASEC and National Health Interview Survey data to understand the reasons for discrepancies in survey reports of health insurance coverage under these programs.

The work on StARS and the other administrative records acquired by the Census Bureau to date represents an excellent start on building the infrastructure to support widespread use of administrative records in Census Bureau programs and in exploring uses of different kinds of records. The bureau's administrative records program, both now and in the future as it adds new sets of records and analysis capabilities, will be an important resource for applications of administrative records in a reengineered SIPP. Beginning with the acquisition of records through data linkage, types of uses, confidentiality protection, and data access, we address issues to consider for SIPP's use of records and outline a goal-oriented approach to identifying the most fruitful applications of administrative records in SIPP for the short and longer terms.

ACQUISITION OF ADMINISTRATIVE RECORDS

The first hurdle for the use of administrative records in a reengineered SIPP is to determine the feasibility and costs of acquiring records from agencies that have custody of them. This hurdle turns out to be much higher for records held by state agencies than for records held by federal agencies.

Federal Records

Through its StARS Program, described in the preceding section, the Census Bureau already has arrangements in place to acquire, update, link, unduplicate, and evaluate information from a large number of administrative records systems from federal agencies. These records provide national coverage for the programs to which they apply. They vary in timeliness. For example, the 2008 StARS file contains the following:

- IRS 1040 records filed any time in 2008, pertaining to 2007 income, which are provided to the Census Bureau in two waves—in October for weeks 1-39 and in January for weeks 40-52.
- IRS 1099 records filed in weeks 1-41 of 2008, pertaining to 2007 income (the bureau does not acquire 1099 records filed in weeks 42-52).
- Medicare Part B enrollment records filed any time in 2008.
- U.S. Department of Housing and Urban Development, Indian Health Service, and Selective Service System records provided to the Census Bureau in May 2008.

SSA files are provided to the Census Bureau with very little delay. The longest time lag is for Medicaid files, which the Census Bureau does not receive from the Centers for Medicare and Medicaid Services (CMS) until 3 years after the reference date.

In addition to the files enumerated above, the Census Bureau is seeking to acquire files from the U.S. Department of Veterans Affairs (VA), and it has access to, but has not used, the Free Application for Student Aid (FAFSA) files from the U.S. Department of Education.[7] The bureau to date has not attempted to, but presumably could, obtain records for Medicare Part D (prescription drug coverage).

The Census Bureau's program to acquire federal administrative records demonstrates a high level of professionalism and competence in negotiating data acquisition and use agreements specific to each provider agency; developing and refining procedures for accurate matching, unduplication, and imputation of missing demographic characteristics; and building systems to enhance the level of confidentiality protection. The research and development work that underlies the StARS Program should greatly facilitate the reengineering process for SIPP, in both the short and longer terms.

In concert with its work to develop StARS and associated records as a Census Bureau–wide resource, we encourage the bureau to systematically outline a plan for acquiring additional federal agency administrative records that are germane to SIPP's goal of providing detailed information on the short-term dynamics of economic well-being for families and households, including employment, earnings, other income, and program eligibility and participation. Acquisition and use of the VA and Medicare Part D files mentioned above should be part of the bureau's plan. Another possibly

[7]The FAFSA files are of limited use because of the limited nature of the population that applies for this aid.

useful source of information is the Federal Case Registry of Child Support Orders maintained by the Administration for Children and Families (ACF) in the U.S. Department of Health and Human Services, although access to this file is difficult to obtain (see http://www.acf.hhs.gov/programs/cse/newhire/fcr/fcr.htm).

The ACS Office of Child Support Enforcement also maintains the National Directory of New Hires (NDNH), which was mandated by the 1996 Personal Responsibility and Work Opportunity Reconciliation Act to assist state child support agencies in locating parents and enforcing child support orders. The NDNH includes quarterly reports from states of new hires in the state (information reported on W-4 forms by employers), quarterly reports of employment and earnings from state workforce agencies (the same data obtained by the Census Bureau's LEHD Program), and quarterly reports from state workforce agencies of unemployment insurance claimants. Federal agencies also report new hires and employment and earnings to the NDNH. The authorizing legislation lists several entities that are entitled to request NDNH information for specific purposes, such as the secretary of education for collection of student loans. The Office of Child Support Enforcement requires a memorandum of understanding and cost-reimbursement for each request of NDNH data. The Census Bureau is not listed as an authorized user of the NDNH; however, "researchers/others" may request NDNH information for research purposes "found by the Secretary of HHS to be likely to contribute to achieving the purposes of Part A or Part D of the Social Security Act" (see http://www.acf.hhs.gov/programs/cse/newhire/library/ndnh/background_guide.htm). The Census Bureau should investigate this source to determine if it could provide employment, earnings, and unemployment benefits information for all states for use in SIPP and other bureau programs without the necessity to negotiate with individual states.

For purposes of a reengineered SIPP, some federal records are more useful than others, not only because they are available with a relatively short time lag, but also because the provisions governing their use are more flexible. In contrast, some federal records, such as IRS records, are very tightly restricted, so that they could be used indirectly but not directly in SIPP. Even for indirect uses, such as evaluation, the available federal records are not a comprehensive resource for SIPP. They do not cover some important sources of income, such as income from some state-administered programs, as well as detailed components of asset income, including dividends and interest by specific asset types (e.g., savings accounts versus money market funds).

State Records

The picture for state administrative records is much less promising. At present, the Census Bureau has records for selected programs for specific years for a few states, including TANF records for Illinois, Maryland, and Texas; food stamp records for Illinois, Maryland, Minnesota, and Texas; general assistance records for Illinois and Maryland; and child care subsidy records for Illinois, Maryland, and Texas. These records have all been acquired for specific research and evaluation purposes (described above). In addition, the Census Bureau has quarterly employment and earnings records from 47 states and the District of Columbia on an ongoing basis through the LEHD Program. The Census Bureau can also in some instances of state-administered programs (e.g., food stamps) obtain counts of recipients by state or county.

To determine the feasibility of acquiring state agency administrative records for a reengineered SIPP, the panel commissioned a study of state laws on confidentiality and access for all 50 states for the TANF, Medicaid, UI, WC, and other cash benefit (principally general assistance) programs. The study was able to find applicable statutes (or determine that the state had no statutes about confidentiality and access for any of these programs) for all 50 states.

The study classified states into three categories (Sylvester, Bardin, and Wann, 2008:5; the category names are the panel's):

1. Ready access—states or state agencies for which the authors could either find (a) *specific* enactments empowering a state agency to provide access to program records for purposes that could include their use for a reengineered SIPP or (b) *no* statute or administrative section that applied to the confidentiality or use of program records.
2. Restricted access—states or state agencies for which the authors could find specific enactments allowing the release of records for purposes that could include their use for a reengineered SIPP but that contain codified restrictions on access, disclosure, or use that the Census Bureau would need to agree to in a memorandum of understanding.
3. No access—states or state agencies for which the authors could find either (a) *general* (constitutional, judicial, or statutory) laws prohibiting access to state-held program records for a purpose such as their use for a reengineered SIPP or (b) *specific* laws prohibiting a state agency from releasing program records for a purpose such as their use for a reengineered SIPP.

Table 3-1 shows how Sylvester, Bardin, and Wann (2008) classified states among categories 1, 2, and 3 for four programs the authors studied—TANF, Medicaid, UI, and other assistance (e.g., general assis-

TABLE 3-1 Classification of State Legal Codes Regarding Access to Records of Four Programs—Medicaid, Temporary Assistance for Needy Families (TANF), Unemployment Insurance (UI), and Other Cash Benefits—for Possible Use by the Census Bureau for SIPP

State	Category 1 Programs	Category 2 Programs	Category 3 Programs
Alabama			All four programs
Alaska			All four programs
Arizona	All four programs		
Arkansas			All four programs
California			All four programs
Colorado			All four programs
Connecticut	UI*		Medicaid, Other cash benefits, TANF
Delaware			All four programs
Florida			All four programs
Georgia			All four programs
Hawaii		UI	Medicaid, Other cash benefits, TANF
Idaho		UI	Medicaid, Other cash benefits, TANF
Illinois		Medicaid, UI	Other cash benefits, TANF
Indiana	Other cash benefits*		Medicaid, TANF, UI
Iowa			All four programs
Kansas	Medicaid*		Other cash benefits, TANF, UI
Kentucky		All four programs	
Louisiana			All four programs
Maine			All four programs
Maryland	UI*		Medicaid, Other cash benefits, TANF
Massachusetts		All four programs	
Michigan	Other cash benefits, TANF	UI	Medicaid
Minnesota		All four programs	
Mississippi		All four programs	
Missouri		UI	Medicaid, Other cash benefits, TANF
Montana		All four programs	

Nebraska	Other cash benefits,* TANF,* UI*		Medicaid
Nevada	Other cash benefits, TANF*	UI	Medicaid
New Hampshire	UI*		Medicaid, Other cash benefits, UI
New Jersey		Medicaid, Other cash benefits, TANF	UI
New Mexico		Other cash benefits, TANF, UI	Medicaid
New York	All four programs*		
North Carolina	UI*	Medicaid, Other cash benefits, TANF	
North Dakota	UI*		Medicaid, Other cash benefits, TANF
Ohio	UI*	Medicaid, Other cash benefits, TANF	
Oklahoma			All four programs
Oregon		UI	Medicaid, Other cash benefits, TANF
Pennsylvania		Other cash benefits, TANF, UI	Medicaid
Rhode Island	All four programs*		
South Carolina	UI*		Medicaid, Other cash benefits, TANF
South Dakota	Other cash benefits,* TANF,* UI*		Medicaid
Tennessee	All four programs*		
Texas	Other cash benefits,* TANF,* UI*		Medicaid
Utah	Medicaid*	Other cash benefits, TANF, UI	
Vermont	All four programs*		
Virginia		Medicaid	Other cash benefits, TANF, UI
Washington		UI	Medicaid, Other cash benefits, TANF
West Virginia		All four programs	
Wisconsin	TANF*	Other cash benefits, UI	Medicaid
Wyoming	UI*		Medicaid, Other cash benefits, TANF

NOTES: Other cash benefits include such programs as general assistance; the categorization of UI also applies to workers' compensation.

Category 1 includes states or state agencies with (a) *specific* enactments empowering a state agency to provide access to program records for purposes that could include their use for a reengineered SIPP or (b) *no* statute or administrative section that applied to the confidentiality or use of program records.

continued

TABLE 3-1 Notes continued

Category 2 includes states or state agencies with specific enactments allowing the release of records for purposes that could include their use for a reengineered SIPP but with codified restrictions on access, disclosure, or use that the Census Bureau would need to accept in a memorandum of understanding.

Category 3 includes states or state agencies with either (a) *general* (constitutional, judicial, or statutory) laws prohibiting access to state-held program records for a purpose such as their use for a reengineered SIPP or (b) *specific* laws prohibiting a state agency from releasing program records for a purpose such as their use for a reengineered SIPP.

* Indicates that Sylvester, Bardin, and Wann (2008) could not find any statute about confidentiality or use of program records.

SOURCE: Classification by Sylvester, Bardin, and Wann (2008:8-9, 12-13, 15-17); panel staff resolved several inconsistent classifications by examining the statutory language provided in Sylvester, Bardin, and Wann (2008:Appendix C).

tance).[8] In total, with 50 states and 4 programs, there are 200 state-program combinations. Of these, the study classified 45 state-program combinations (in 22 states) in Category 1 (ready access)—because the state either explicitly permits or does not prohibit the use of program records by an agency such as the Census Bureau for statistical purposes (mainly the latter). The study classified another 42 state-program combinations (in 18 states) in Category 2 (restricted access) because the state would permit the use of program records by the Census Bureau under more or less restricted conditions. Finally, the study classified 113 state-program combinations (in 38 states) in Category 3 (no access) because the state generally or specifically prohibits the use of program records by an agency such as the Census Bureau for statistical purposes.

The classifications in Table 3-1 represent the authors' judgments based on their review of state constitutions and legal codes, excluding regulations, executive orders, and other possible kinds of interpretations that might allow access to records by the Census Bureau for statistical purposes. Indeed, members of the panel are aware of instances in which records from some of the states listed in the "no access" category have been used for research, although some of these instances may have applied to records not covered by Sylvester, Bardin, and Wann (e.g., food stamps) and to uses within the state and not by a federal agency.[9]

Nonetheless, it would appear from the analysis of Sylvester, Bardin, and Wann (2008) to be impossible for the Census Bureau to acquire records for all programs of interest for all 50 states and difficult for it to acquire records for more than a handful of states. Not only do 38 states in their analysis apparently preclude access to records for at least 1 of the 4 programs studied (Category 3), but another 18 states place restrictions on access (Category 2). Some of the legislative provisions for Category 2 states are relatively benign, such as requiring access to be "in the public interest" or for official purposes. The legislative provisions for other states in this category are more onerous, such as requiring advance notification and consent from individuals in a program. Finally, of the 22 states that would appear able to provide records to the Census Bureau for a reengineered SIPP for at least 1 of the 4 programs studied (Category 1), only 2 states (Arizona for all 4 programs and Michigan for TANF and other cash benefits) have statutes

[8] The categorization shown in Table 3-1 for UI also applies to workers' compensation—Sylvester, Bardin, and Wann treated UI and WC as a single program in their review because both programs are administered by the same office in each state.

[9] For example, Hotz, Mullin, and Scholz (2003, 2005) have analyzed matches of California public assistance, unemployment insurance, and tax records, but the matches were performed by the California Tax Franchise Board, which delivered aggregated results to the researchers. The ADARE alliance provides access to state records by authorized researchers but not by federal agencies.

that explicitly allow for data sharing with a federal agency. The remaining states are in Category 1 because Sylvester and his coauthors could not find any statutes pertaining to confidentiality and access, yet such states may well have regulations that limit access.

The situation is far from hopeless, however. One pattern that emerges from the data collected by Sylvester and his coauthors is that access to UI records—and perhaps WC records—may be possible in many states, either because of a statute that permits access (albeit often with restrictions) or because there appears to be no applicable statute that would prohibit access. Moreover, many states are statutorily allowed to provide records to other states or even federal agencies for purposes of program administration. Although the Census Bureau is not a program administration agency, the data from SIPP could be useful to states for program evaluation and improvement. Just as the Census Bureau provides quarterly workforce indicators in return for access to state employment security agency employment and earnings records for the LEHD Program, it might be possible to develop an appropriate quid pro quo that would benefit state agencies that provide records for a reengineered SIPP (see "Strategic Planning for Acquisition" below).

Finally, it is important to note that the distribution of program benefits is not uniform across the states, which means that coverage of a significant proportion of the caseload for such programs as TANF and food stamps could be obtained by acquiring records from a relatively small number of states. For example, the TANF records for the two states of Illinois and Texas currently available to the Census Bureau cover about 8 percent of TANF recipients nationwide. If it were possible to acquire TANF records from just five more states, including California, Michigan, New York, Ohio, and Pennsylvania, coverage could be extended to one-half of TANF recipients nationwide, greatly facilitating indirect uses of administrative records in SIPP, such as evaluation and improved imputation procedures.

Strategic Planning for Acquisition

We applaud the Census Bureau's work on acquiring federal administrative records, which have great potential value for a reengineered SIPP in addition to many other bureau programs. The Census Bureau should continue that work and seek to acquire additional federal records to the extent possible, such as VA records. For federal records, the costs of acquisition, matching, and editing appear to be low compared with the benefits and have the advantage that they can be spread over many Census Bureau programs.

In contrast, the costs for the Census Bureau in attempting to acquire and use state program records would be substantial. These costs would include the time and effort to make contact with appropriate state agencies, verify the provisions of state statutes and regulations that pertain to confidentiality and data access, and develop acceptable memoranda of understanding. In addition, there would be costs, subsequent to acquiring records, to clean and edit the data, which would probably necessitate time-consuming interactions with state agency staff, or with research organizations that are knowledgeable of the state files, to answer questions and resolve discrepancies. Moreover, some attempts to acquire records would be likely to come to naught, even with the expenditure of substantial time and resources to develop a mutually acceptable memorandum of understanding for data acquisition.

Given these challenges, the Census Bureau will need to think strategically about acquisition of state records and develop a well-thought-out plan for acquisition in the short and longer terms. By "think strategically," we mean that the Census Bureau will need to develop priorities for acquisition of state records in light of the goals of SIPP and the importance of different kinds of program records for those goals. Three criteria for establishing priorities include the importance of the income source for lower income households, particularly in times of economic distress; the relative ease of acquiring the records; and the ability to cover a large proportion of the program caseload by acquiring records from a relatively small number of states.

As an example, consider UI benefits. Subsequent to the enactment of welfare reform in 1996, more low-income single mothers with children entered the workforce and so were able to turn to UI benefits when they lost a job. By 2002, more single mothers with children received UI than TANF benefits (Assistant Secretary for Planning and Evaluation, 2005:Figure C). Such findings, coupled with the importance of being able to analyze the contribution of UI benefits to ameliorating recessionary economic conditions and the fact that UI records may be easier to obtain than other kinds of state records, suggest that UI records could be a target of opportunity for the Census Bureau. Moreover, the relationships built by the LEHD Program with state employment security agencies may facilitate obtaining not only employment and earnings records, as is done in the LEHD Program, but also UI records. It may also be possible, as noted above, to acquire UI records for all states from the federally maintained National Directory of New Hires, which could be an efficient, low-cost source for acquiring these data, providing the Census Bureau could obtain permission to use the data for improving SIPP.

In contrast, a program such as WC contributes less to aggregate income than the UI Program. Moreover, duration of benefit receipt tends to be longer, while aggregate amounts of benefits paid show no particular trends over time. These factors suggest that obtaining WC records is of lower priority for SIPP's primary purpose of supporting policy analysis and research on intrayear dynamics of program participation and income. Of course, it if were readily possible to acquire WC records at the same time and under the same provisions as UI records, the Census Bureau should not hesitate to do so.

In addition to setting priorities among program records for acquisition, the Census Bureau will need to take account of acquisition issues in determining the types of uses to which it will put the records it acquires. In the short term, promising to restrict use of records to indirect uses, such as evaluation and perhaps improvement of imputation methods, could facilitate acquisition because the threats to confidentiality would be substantially lower than if the records were to be used directly in a reengineered SIPP. In the longer term, it may be possible to move toward direct uses once ongoing relationships have been built with state agencies and by developing ways to provide states with useful information, as has been done in the LEHD Program. For example, sample size might be added for states that are very cooperative about providing program records, so that the SIPP data for those states would be statistically reliable for analysis at the state level. Adding sample could significantly increase SIPP's costs, but there could be substantial benefits of higher quality data given that the survey historically produces net underestimates of many sources of income.

Overall, the reengineering of SIPP will need to proceed on the assumption that significant use of state administrative records cannot be part of the plan in the short and medium term. Nevertheless, a strategy for the acquisition of high-priority types of state records and their use for such purposes as evaluation of SIPP data quality should be developed and implemented as resources permit. In addition, the reengineering plan should envision a wide variety of uses of federal records.

LINKAGE OF ADMINISTRATIVE AND SURVEY RECORDS

Many applications of administrative records in a reengineered SIPP require matching of the administrative and survey data for individuals and households. Fellegi and Sunter (1969) provided the first formal mathematical model for probabilistic record linkage techniques, building on ideas introduced by Newcombe and colleagues (1959). Beginning in the late 1980s, Census Bureau staff have been leaders in the development and continuous improvement of computer-based record linkage software of the kind that underlies the StARS database (see Winkler, 2006, for a review article on research and development in the record linkage field). Census Bureau staff and others have addressed such challenges as standardizing names and addresses across data files to reduce rates of false negatives (failure to find a match when one exists); developing algorithms to compare strings of characters (e.g., names) among data files that allow for typographical errors in one or both files (even after standardization) being matched; forcing one-to-one matches to reduce rates of false positives (matching two records that are not for the same individual); developing methods to block or group records in ways that make the searching and matching processes more efficient; and developing methods to use auxiliary data files to improve the match between two files.

The Census Bureau clearly knows how to conduct efficient, high-quality matches of data files, even when SSNs are not available, as has been the case with SIPP responses since about 2006 (midway through the 2004 panel). While never perfect, such matching has been shown to achieve good results. For example, as noted above, a match of the 2001 March CPS with the Numident file using the person verification system was successful 94 percent of the time using SSNs (available for about 53 percent of the CPS records) and 93 percent of the time using only name, address, and date of birth. (The universe for matching excluded refusals and records lacking a name.) Extensive review estimated the false match rate to be very low—between 0.13 and 0.20 percent. The estimated false nonmatch rate was higher—4.65 percent (Wagner, 2007).

Another evaluation compared the demographic composition of records from the 2001 ACS that matched and did not match the Numident file on the basis of name, address, and date of birth (SSNs are not collected in the ACS). The matched cases (91 percent of the total eligible for matching) were very similar in distribution by gender, race, Hispanic origin, age group, and income group to the full ACS file. The not-matched cases (9 percent of the total) differed significantly in composition: Compared with the full ACS file, the not-matched cases included higher proportions of minorities, younger people, and lower income groups. These results could reflect not only that minorities, younger people, and lower income groups are less likely to have SSNs, but also that the information on name, address, and date of birth for these groups is more likely to differ between the Numident and other files.

Matching errors should not be ignored, particularly false negatives that underestimate the true match rate and negate the possibility of using administrative records for people who should be but are determined not to be a match. However, the error rates evident in the evaluations of which we are aware appear to be smaller than the missing data rates that surveys often experience in reports of income, employment, and other characteristics. We encourage the Census Bureau to view the errors in administrative records and in matches of them with survey records in the same manner that the bureau and other statistical agencies have commonly viewed nonresponse and reporting errors in surveys—namely, as problems to address but not a brick wall. Some of the same techniques that are used to evaluate survey reporting errors, such as reinterviews of samples of respondents and efforts to track down nonrespondents, could well be applied to evaluating and perhaps correcting data quality problems with administrative records and matching.

INDIRECT USES

We now come to the question of the kinds of uses that administrative records can play in a reengineered SIPP. We begin with indirect uses, in which the data from administrative records never replace or add to the data in SIPP public-use microdata files. The advantage of indirect uses of administrative records is that they do not increase (in the case of evaluation), or only minimally increase (in the case of their use in imputation models), the risk of identification of SIPP respondents in public-use files. Consequently, these uses do not necessitate much if any in the way of additional confidentiality protection procedures. The disadvantage is that indirect uses of administrative records may not improve data quality to the extent possible with direct use.

Aggregate Comparisons

The history of SIPP's uses of administrative records outlined above notes several examples of using aggregate estimates from administrative records to evaluate corresponding aggregate estimates from the survey, such as aggregate benefits received from an assistance program or average monthly participation in a program. This use of administrative data is relatively inexpensive; the major difficulty lies in making appropriate adjustments to the administrative data estimates or survey estimates or both to make them as comparable as possible with regard to the universe of people covered, the time period covered, and the definition of participation and income. Also, this use of administrative data, given that comparisons are at the aggregate level, is only the starting point of work to evaluate and improve the quality of the survey data. Yet aggregate comparisons are an important first step, one which we think the Census Bureau should put on a regular schedule and routinize to the extent possible.

The reason that aggregate comparisons should be made on a regular basis is evident from examining some of the comparisons that have been performed. For example, Meyer, Mok, and Sullivan (2009:Table 2) found that SIPP estimates of aggregate dollar benefits from AFDC and its successor TANF as a ratio of program estimates have fluctuated over time, with a pronounced downward trend beginning in 1998. In contrast, SIPP estimates of average monthly participation in AFDC/TANF have not shown a time trend up or down (Meyer, Mok, and Sullivan, 2009:Table 11). These disparate findings suggest avenues of research for the Census Bureau to explore, such as evaluating individually matched records for the states that have provided them to the bureau and engaging in questionnaire design research to try to make the reporting of benefit amounts at least as accurate as the reporting of program participation.

Most aggregate-level comparisons have to be made at the national level for the population as a whole given the limitations of available data. Of course, when the Census Bureau has access to 100 percent of program records, as in the case of such federal programs as SSI, it can perform comparisons at any level of aggregation that is desired, including at the individual record level (see "Individual-Level Comparisons" below).

For state-administered programs, it may be possible in some instances to obtain more disaggregated estimates for comparison. For example, the Food and Nutrition Service provides state and county counts of monthly food stamp recipients to the Census Bureau for its SAIPE/SAHIE programs. These estimates could be used to develop ratios of monthly participants in SIPP versus the monthly program counts by geographic areas that could illuminate differences in reporting patterns that warrant research. For example, ratios of SIPP reporting to administrative totals in central-city counties (e.g., Chicago, Los Angeles) may differ from the ratios in suburban and rural counties. The SIPP data would need to be combined to form groups of counties for which SIPP estimates were sufficiently reliable for comparison purposes. (Aggregating more than 1 year of data could be helpful in this regard.) The Employment and Training Administration in the U.S. Department of Labor makes available weekly counts of unemployment benefit claims by state, which could be analyzed in a similar fashion (see http://workforcesecurity.doleta.gov/unemploy/finance.asp).

Some state-administered programs, such as TANF, food stamps, and unemployment insurance, also provide periodic reports to the relevant federal agencies on characteristics of benefit recipients, most often drawn from samples of state administrative records (see, for example, http://aspe.hhs.gov/HSP/alt-outcomes00/app_d.htm and http://workforcesecurity.doleta.gov/unemploy/chariu.asp). These statistics could be useful for comparison purposes, although they would be subject to sampling error and also nonsampling error, in that reports of characteristics of program caseloads, such as other sources of income, may be less accurate than benefit amounts. Nonetheless, some characteristics in the administrative statistics, such as type of TANF recipient unit—single adult and children, two-parent family, or children only—may be deemed accurate enough to be useful for comparison with SIPP estimates.

To facilitate a program of regular aggregate comparisons, which should include not only SIPP, but also the CPS and perhaps other surveys that ask about income and program participation, the Census Bureau should explore with the Office of Management and Budget Statistical and Science Policy Office the establishment of an interagency technical working group to support the effort. Staff from such agencies as the Administration for Children and Families (which oversees TANF), the Food and Nutrition Service (which oversees the food stamp, school meal, and WIC programs), the Internal Revenue Service (which oversees income reported on tax forms), and other agencies could be detailed to work with Census Bureau staff to develop the most comparable estimates possible for their programs. In this way, aggregate comparisons could be prepared on a recurring basis that would make use of the program knowledge in the agencies and the survey research knowledge in the Census Bureau to ensure the highest quality and most useful comparisons. Such comparisons, regularly disseminated, should be very useful to policy analysts and other data users in the public and private sectors. The members of the interagency technical working group could also contribute to the use of administrative records for other purposes, such as evaluating and improving imputation models for missing data. (See Chapter 4 for a related recommendation on obtaining assistance from researchers and policy analysts with regard to aggregate comparisons, imputation models, and other applications of administrative records.)

Individual-Level Comparisons

In addition to aggregate comparisons, individual-level comparisons of matched administrative and survey records are important to carry out because they make it possible to estimate the extent of gross errors—that is, overreporting and underreporting—whereas aggregate comparisons make it possible to estimate only net errors. Individual-level comparisons can shed light on whether reporting errors are random or systematic, and, if the latter, whether they relate to other characteristics of respondents in ways that could suggest improved questionnaire design or other aspects of a survey. Examples of systematic error are the confusion among Social Security and SSI benefit receipt found by Huynh, Rupp, and Sears (2001) and also their finding that imputed benefits are much less accurate than reported benefits. If the gross errors for an income source are very large, then that may suggest giving serious consideration to using the administrative data to correct the survey reports.

For evaluation of income sources and program participation for state-administered programs, it is not necessary to acquire records for all or a large proportion of states in order to generate useful findings. The comparisons currently under way of TANF and food stamp reporting from administrative records with SIPP survey and event history calendar reports for 2007 for a subsample of the 2004 SIPP panel in the two states of Illinois and Texas should yield useful findings that suggest further avenues for fruitful research. (See Chapter 4 for a discussion of the limitations of the comparisons with the event history calendar reports, which are paper based.)

Resources permitting, the Census Bureau should not stop with the Illinois-Texas comparisons for TANF and food stamps—and, indeed, the bureau is endeavoring to obtain administrative records from other states for use in evaluating the results of its electronic event history calendar test in early 2010 (see Chapter 4). Working from a strategic plan, developed in consultation with SIPP data users, which considers the importance of an income source for low-income households and the feasibility of acquiring records for a significant proportion of program participants, the Census Bureau should identify priority programs and states to pursue for the purpose of acquiring records under mutually acceptable memoranda of understanding.

In addition to following a targeted strategy for the acquisition of selected state records, the Census Bureau should carry out individual-level evaluations for federal records that it already holds as part of its StARS database and for the state records of employment and earnings that it acquires for the LEHD Program. Again, the bureau should plan strategically for which programs to evaluate in the short and longer terms.

Use of Administrative Records in Weighting

Like other surveys, SIPP assigns weights to each person in the sample so that estimates from the data, obtained by applying the appropriate weights, represent the survey universe. The Census Bureau provides cross-sectional and longitudinal (panel) weights on SIPP data records to facilitate different uses of the data.

SIPP weighting routines, as in other Census Bureau surveys, not only make use of the inverse of the sampling probability and adjustment factors for whole-household nonresponse, but also include adjustment factors to bring estimates for age, gender, and race and ethnicity categories into agreement with independently estimated population control totals for these groups. The use of population controls is essential in the weighting process because without them the survey would significantly underrepresent important demographic groups, such as young minority men (see, e.g., U.S. Census Bureau, 1998:Tables 3-4, 3-5, 3-6).

The Census Bureau develops population controls from the decennial census updated with administrative records on births, deaths, and net international migration. However, demographically based controls do not take account of other characteristics that may distinguish well-represented from underrepresented groups in the survey. In this regard, we encourage the Census Bureau to revisit its earlier research on using IRS tax record data in SIPP weighting to reduce the variance of income estimates to see if that research could be worth pursuing for a reengineered SIPP (see "SIPP's History with Administrative Records" above).

Improving Imputations

Imputation Methods in SIPP

SIPP, like other surveys, has missing data, which the Census Bureau processes so that the resulting data file represents the population that was sampled and has values for every item for every person and household in the file. There are three main types of missing information:

1. Whole-household nonresponse, which is handled by a nonresponse adjustment in the calculation of weights for the responding households.
2. Partial household nonresponse, in which a member of an otherwise responsive household fails to respond or provides too few items of information. Called Type Z noninterviews, these cases are typically handled by a procedure in which the entire record of another respondent that is similar to the nonrespondent on demographic characteristics that are available for both is substituted for the nonrespondent.
3. Item nonresponse, in which a respondent answers some but not all questions. Values for missing items are supplied through edits based on other information in the person's own record or, more often, from hot-deck imputation, which also is used for some Type Z noninterviews.

Hot-deck imputation for item nonresponse has a long history at the Census Bureau, beginning with the 1960 decennial census (see National Research Council, 2004b:458-459), and is widely used in the bureau's census and survey programs. To explain, but oversimplifying: the records in a data file are sorted, usually by geographic area of residence; valid responses for a variable are continually entered into the cells of an appropriate imputation matrix as the data file is processed; and the most recent (hottest) valid value is substituted for a missing response. The geographic sort helps ensure that responses are imputed from a person living in the same or nearby area. The imputation matrix for a variable or a collection of related variables usually includes demographic characteristics, such as age category, gender, race, and ethnicity, and may also include other variables. The intent is to supply a hot-deck value from a donor record that is very similar to the respondent; when this is not possible, the matrix categories are collapsed as necessary to find a donor. As a last resort, the starting, or cold, value, for the variable, which is prespecified, is used to supply a response.

A problem with the hot-deck method, as it has been employed for SIPP, is that the variables that define the categories in a particular matrix are often not carefully tailored to the variable being imputed. Without careful tailoring, program participation, for example, may be imputed to people whose incomes from other sources would render them ineligible to receive benefits, or, alternatively, too high income amounts for, say, wages or property income may be imputed to people who report that they are participating in a means-tested assistance program (see Appendix A; see also McKee and McBride, 2008). Yet the more variables that are included in the matrix, the harder it may be to find a donor, and the more often that a single record may be used to supply values for large numbers of records with missing responses. Collapsing matrix cells provides more donors but at the cost of greater heterogeneity of the donor pool.

Model-Based, Multiple Imputations

The Census Bureau could better handle missing data in SIPP with modern, flexible model-based imputation techniques, which take account of more information than the hot-deck method. In fact, bureau staff are beginning research on model-based approaches for SIPP imputation (Stinson, 2008).

To illustrate how a model-based approach might be useful in SIPP, we suppose there are missing values of program participation status (assumed to be a binary indicator variable) for only one particular month; no other variables are missing. To handle missing values for multiple variables simultaneously, the Census Bureau can use the multivariate imputation approach of Raghunathan et al. (2001). This approach relies on a collection of imputation models for each variable with missing values, so that the general principles for the one variable scenario are useful for multivariate scenarios.

The first step of the process is to fill in any missing values that are determined by program rules. For example, if program participation is contingent on income not exceeding some threshold, all people whose incomes exceed that threshold are imputed to be nonparticipants (i.e., status = 0).[10]

[10]Program eligibility rules, in practice, are more complicated than a simple income threshold; they may involve not only income level, but also family composition, citizenship status, the value of certain types of assets, work expenses, out-of-pocket medical care expenses, shelter expenses, and the like. To the extent the Census Bureau can mimic the eligibility rules for a particular program in an imputation model, the better; however, applying even a simple income threshold is preferable to allowing program participation to be imputed to any record with a missing value.

Or, if participants in one program, such as SSI, cannot also participate in another program, such as TANF, all people reporting, or imputed, to receive SSI would be imputed to be nonparticipants in TANF. Such checks can be automated in an imputation software routine.

The second step is to impute values for people eligible for participation (i.e., status $\neq 0$). To do so, the Census Bureau could estimate a logistic regression of the participation status indicator on predictors associated with program participation. Only records eligible for participation are used to fit the regression. The predictors might include demographic variables such as age and gender, economic variables such as income, participation status from other months and other programs, and even data from other waves of SIPP. In general, it is prudent to include all variables thought to be associated with participation status, as this improves the chances that important relationships will be preserved in the completed data sets. If the Census Bureau suspects that the regression coefficients differ by population group, it could split the sample by these population groups and estimate the regression separately for each. Once the model is estimated, the Census Bureau would compute the resulting predicted probabilities and randomly sample missing participation status values from Bernoulli distributions with these probabilities. In addition, proper imputations would also use Bayesian methods to account for the uncertainty in the predicted probabilities. Standard imputation software incorporates this uncertainty automatically.

If using model-based imputation, the Census Bureau should strongly consider creating multiple imputations rather than single imputations for each missing datum. Multiple imputations allow users to incorporate estimates of the uncertainty introduced by imputation into calculations of standard errors, by using standard complete-data methods and simple rules for combining estimates from the multiple data sets. For details on the benefits of multiple imputation, see Rubin (1987) and Schafer (1997). For examples of the use of multiple imputation in large federal surveys, see Schenker and colleagues (2006), which describes multiple imputation of income and earnings data in the National Health Interview Survey (see also Parker and Schenker, 2007), and Kennickell (2006), which describes multiple imputation of assets and liabilities data in the Survey of Consumer Finances, which was implemented when the survey was redesigned in 1989.

Use of Administrative Records in Model-Based Imputations

As Stinson (2008:7) notes, "all imputation methods that use survey data exclusively are built on the assumption that the relationships between survey variables are the same for everyone, regardless of missing data." This is the "missing at random" (MAR) assumption. However, if the relationship between a variable such as program participation and variables that are predictive of participation differs when program participation is not reported, then an imputation that uses survey data alone will be flawed.

Administrative records could be used to evaluate and improve model-based imputations in this regard.[11] For example, the Census Bureau recently conducted an evaluation of earnings responses and imputations in the 2004 SIPP panel compared with earnings information reported on W-2 records to which it has access from IRS (Stinson, 2008:9-13). For this evaluation, the Census Bureau divided SIPP respondents into 4 groups on the basis of the number of months in which earnings were imputed for one or more jobs reported for calendar 2004: no months of imputed or missing data; 1-4 months of imputed data; 5-8 months of imputed data; 9-12 months of imputed data. Regressing the W-2 earnings on SIPP demographic characteristics for each of the four groups, predicting earnings for each group using the coefficients from each of the four regression equations, and averaging the differences of the W-2 earnings from the predicted earnings should give results of about zero for each group if the missing data are MAR. However, the evaluation results indicated that for Group 2 (1-4 months imputed data), the imputed earnings appear to be too high on average, while for Group 4 (9-12 months of imputed data), the imputed earnings appear to be too low. Similarly, the work cited earlier by Huynh, Rupp, and Sears (2001:Table 7) documented that the current hot-deck model for imputing SSI and OASDI benefits does not do a good job—it imputes benefits that are too high, on average, compared with program records, particularly for SSI, indicating that nonrespondents differ from respondents in ways that are not captured in the hot-deck matrix.

On the basis of these kinds of evaluations, the Census Bureau could profitably revise its imputation models to include administrative records in order to improve the accuracy of the imputed values. An advantage of this use of administrative records is that timely availability of the records would not be critical. Presumably, imputation models would be developed on the basis of the most recent data available and reestimated with newer data only every few years.

Specific ways in which SIPP imputation models could use administrative records will vary, depending on such factors as whether the Census Bureau has access to a particular set of records nationwide or only for some states, whether it has access to the individual administrative records or only to aggregated information, and whether participation or benefits or both together are being imputed. As just one example, consider a federal

---

[11]Using administrative records in hot-deck imputation matrices, while possible, does not make sense unless the administrative values are also used to substitute for or adjust the survey responses; otherwise the imputed values will be inconsistent with the reported survey values given the net underreporting of participation and benefits for many programs.

program such as SSI, for which the bureau has access to 100 percent of the records from the SSA. Instead of a hot-deck imputation, the Census Bureau could match the survey and SSI program records and then develop a model to jointly predict actual SSI participation and benefit amounts from characteristics reported in SIPP. Because there is relatively little net underreporting of SSI participation or benefits in SIPP (even though respondents often confuse SSI with OASDI receipt; see Huynh, Rupp, and Sears, 2001: Table 2), there would be no need to adjust the predictions from the imputation model for consistency with the actual SIPP reporting, as might be the case for income sources for which there is significant reporting bias.

For state-administered programs, the same kind of modeling of participation and benefits could be done as described for the federal SSI program, except that the modeling would likely be limited to only a few states given the difficulties described above in gaining access to state records for Census Bureau use. Use of a model developed on a subset of states to impute missing values for other states would have to be undertaken with care because of differences among state program rules and policies. For programs with important state variations, the use of an imputation model developed from selected states would probably not be desirable.

In developing, evaluating, and improving model-based imputations in these and other ways, the Census Bureau should be guided by a strategic plan that prioritizes its work according to such criteria as the importance of the income source for key population groups, such as lower income people and the elderly, and the feasibility of acquiring records. In addition, as part of an ongoing program for acquiring and using administrative records in a reengineered SIPP, the Census Bureau should establish a schedule for periodic reevaluation and improvement of model-based imputation routines with administrative records. Imputation routines should not be frozen for years and decades at a time, as has happened historically with SIPP. They should be revisited as records become available from more sources for evaluation purposes (e.g., from additional states) and as programs and economic conditions change in ways that suggest the need to revise one or more imputation models.

Confidentiality Concerns

Including imputations in a file that incorporate information from administrative records introduces far fewer risks to confidentiality protection than does direct substitution of actual values (see "Direct Uses" below). An intruder—namely, someone who tries to reidentify individuals in the file by matching with other data sources, such as data available on the web—cannot be certain that matches based on imputed values are true, since the imputed values are predicted ones that are not necessarily the true values.

In general, for hot-deck or other single imputation strategies, the Census Bureau should compute disclosure risks (refer back to Box 3-1) using the SIPP records both before and after imputation. Using the incomplete records (i.e., with missing responses) mimics an intruder who does not trust the imputations and bases matches only on the values known to belong to the data records. Using the completed records (i.e., with imputed values) mimics an intruder who matches whatever values are released. For multiple imputation strategies, the Census Bureau should match on each completed data set (in multiple imputation, some number, $m > 1$, data sets are released) and average the risk measures across the data sets, as well as quantify disclosure risks based on just the incomplete data.

These kinds of analyses will help the bureau determine whether its current confidentiality protection procedures for SIPP public-use microdata files are unnecessarily stringent, are about right, or need to be enhanced. When administrative records are used to inform model-based multiple imputations, comparisons of disclosure risks for multiply imputed files with and without input from administrative records would indicate what additional confidentiality protection, if any, might be needed for the models that incorporate administrative records.

DIRECT USES

Direct uses of administrative data are uses in which administrative data are incorporated to a greater or lesser extent into survey records. These kinds of uses include providing values directly for missing survey responses; adjusting survey responses for net underreporting or overreporting; using administrative records values in place of asking one or more survey questions; and appending administrative records values to survey records.

Direct uses of administrative records raise confidentiality concerns, which, in turn, could make it more difficult for the Census Bureau to release useful public-use microdata files. Such uses also raise concerns about the possible effects on timeliness and survey response. Administrative records may not be available on a schedule that permits their inclusion with the corresponding survey data on a timely basis. Moreover, most direct uses of administrative records, in contrast to indirect uses, would require that SIPP respondents be informed about such uses, which could increase refusal rates for the survey. Experience with 2004 SIPP panel respondents suggests that the effect on response rates might be small, but that would need to be tested more fully. The custodial agency would also have to agree to direct uses of their records in a reengineered SIPP.

On the positive side, direct uses of administrative records promise significant improvements in the quality of SIPP estimates of income and program participation. Moreover, should the use of annual interviews with an event history calendar for eliciting intrayear information on employment, income, family composition, and program participation prove significantly less effective than desired (see Chapter 4), it could be important to consider ways to use administrative records directly. Otherwise, the quality of data on intrayear dynamics of change would be impaired unless SIPP continues to interview respondents every 4 months and forgoes the cost savings from moving to an annual interview schedule.

Replacing Missing Survey Responses with Values from Records (Direct Imputation)

For income sources and programs for which the Census Bureau has access to administrative records, they could be used to supply values for missing survey responses on a one-to-one basis—that is, an individual's record of participation and income amounts would be matched with and directly entered onto his or her SIPP record without using any type of imputation procedure. An imputation model would be used only for people who did not match to an administrative record or, in the case of state records, for people in states that did not provide records to the Census Bureau.

This use of administrative records seems obviously preferable to model-based imputations that incorporate administrative records, in that the actual values are by definition more accurate than any imputation model could be. Yet direct imputation also raises concerns about the possible adverse effects on timeliness, consistency of reports, and disclosure risk for the resulting public-use microdata files. Direct imputation further assumes that not only the survey respondents but also the cognizant custodial agency officials have agreed to the use of records for this purpose.

Timeliness is a concern with direct imputation because directly replacing missing responses with actual values requires records that relate to the survey reference period and are available soon enough after it that SIPP processing and data release are not delayed. As discussed above, some records are available on such a schedule, and others are not.

Consistency with survey reporting is an issue given the reporting error in SIPP, which most often results in a net underreporting bias. It would be incongruous to have the survey responses reflect biased reporting and the imputed values reflect unbiased reporting, yet it is not clear how to address this problem unless administrative records are also used to adjust the survey reports for net overreporting (or underreporting), as discussed in the next section.

Direct imputation must increase the risk of disclosure compared with the use of administrative records in an imputation model unless additional disclosure protection steps are taken. Not only would an intruder know that the imputed values are the administrative records values, but also an intruder could be someone in the custodial agency with access to and knowledge of specific administrative records values. The increased risk would be lessened to the extent that directly imputed values are adjusted for consistency with the survey reporting, assuming that the adjustment is done stochastically and not in a manner that would be transparent to an intruder (e.g., a simple ratio adjustment). If adjustment for consistency with the survey reporting is not needed, then some kind of probabilistic perturbation of the directly imputed values would probably be required to provide sufficient confidentiality protection, in addition to agreements with the custodial agency that include penalties for a breach of confidentiality by that agency's employees similar to the penalties that are already included in Title 13 of the U.S. Code for Census Bureau staff (see http://uscode.house.gov/download/pls/13C1.txt).

Adjusting Survey Responses

The evidence of net underreporting of participation and benefit amounts in SIPP (and other surveys) for most income sources suggests that it could be desirable to adjust the survey responses for groups of respondents with similar characteristics so that estimates for the total population and population groups approximate estimates from administrative records. Major microsimulation models that federal agencies use for tax and transfer program policy analysis regularly simulate program eligibility, participation, and benefits on such surveys as the CPS and SIPP. The estimates from the models are therefore much closer to administrative records aggregates than the unadjusted survey estimates. For example, Wheaton (2007) reports on work to compare reporting of food stamps, Medicaid, SSI, and TANF in the CPS with estimates from the Transfer Income Model, version 3 (TRIM3), which produces an adjusted CPS. The TRIM3 estimates show a much greater effect of food stamps, SSI, and TANF in lifting program recipients out of poverty compared with the survey estimates (Wheaton, 2007:Tables 4-5). Because SIPP achieves more complete reporting of SSI and food stamps than the CPS (Wheaton, 2007:Table 1), the effects would not be as pronounced for a comparison based on SIPP. Nonetheless, they could still be significant overall and for particular programs, such as TANF, for which estimates from both surveys fall markedly short of administrative records.

The approach the Census Bureau would use to adjust survey reports might not be that used by a microsimulation model such as TRIM3. These models not only adjust reported program benefit amounts for individuals that report recipiency on the survey, but also "create" new recipient units and associated benefits from households simulated to be eligible that did not report participation in order to better approximate administrative aggregates for program caseloads and benefit dollars. The Census Bureau might not want to alter the SIPP records to that extent.

As an alternative, the Census Bureau could follow a three-step process to achieve the same effect:

1. The first step would be to implement model-based imputations of the type described above, in which the model predicts administrative records values for respondents with missing data on program participation and benefits.
2. The second step would be to develop adjustment factors to bring the benefit amounts for respondents who report participating in a particular program up to the same percentage of total dollars as the percentage of reporters is of the total caseload (from administrative records).
3. To account for the remaining underreporting, the third step would be to adjust the survey weights—increasing the weights of respondents who report or are imputed participation and decreasing the weights of other respondents with incomes below a specified threshold that approximates the threshold for program eligibility. A threshold is used so that higher income respondents are not downweighted. Given multiple program participation, there could be a need to adjust the weights for participants in a single program separately from those in multiple programs.

Before implementing such an approach, it would need to be carefully evaluated, in general and for particular programs and combinations of programs, and taking account of the effects on other possibly related variables.

Methodological Considerations

The use of administrative records to adjust survey reports in the manner described requires a high degree of accuracy in achieving comparability of both sources with regard to the population covered and the definition of participant units and income and benefits. With regard to timeliness, it would be important to have as up-to-date administrative information as possible for programs and income sources for which participation is growing (or decreasing) rapidly. For programs and income sources for which growth is more predictable, it could be possible to use a simple time trend factor to update older administrative data for use in adjusting more recent survey reports.

For programs for which the Census Bureau has access to the administrative records, adjustments could be made for finely stratified groups. For programs for which the Census Bureau has access only to aggregate statistics, the adjustments would necessarily have to be made for broad groupings. For state-administered programs for which the Census Bureau has access to records for some but not all states, a combination of records and aggregate statistics by state could be used to compute adjustment factors.

With regard to disclosure risk, the development of adjustment factors to achieve approximate agreement with administrative records should not pose any increased threats to confidentiality beyond those described above in the discussion of using administrative records in model-based imputations. The adjustment factors would pertain to groups and not to individuals.

Strategic Considerations

Adjusting survey responses for net reporting error would lead the Census Bureau in a direction that it is not often accustomed to taking for household statistics—namely, that of producing a set of best estimates by combining sources of information in contrast to producing the data reported from a survey. The Census Bureau produces a small number of model-based estimates in its SAIPE and SAHIE Programs that use both survey and administrative data, but, in each case, the variable predicted is an estimate from a survey, such as the ACS estimate of poor school-age children. Coming closest to the idea of producing the best estimates is the Census Bureau's regular practice of adjusting survey weights so that the survey estimates agree with independent population control totals, which themselves are developed from the previous census updated with administrative records on births, deaths, and net international migration. Another example is the Census Bureau's seasonal adjustment models (see http://www.census.gov/srd/www/x12a/), which it uses to adjust economic time series from business data, and which the Bureau of Labor Statistics uses to adjust monthly unemployment rates from the CPS.

A second consideration that could give the Census Bureau pause about the wisdom of adjusting survey reports is the sheer complexity of the adjustment process, as outlined above, for the large number of programs and other sources of income, such as earnings, dividends, and interest, that could be candidates for its use. As we recommend throughout this report, the Census Bureau, with input from the user community, would need to take a strategic approach in moving toward a goal of adjusting survey responses. It would need to decide which income sources would be feasible to adjust and which would be most important to adjust in terms of the potential effect on research and policy analysis results. Undoubtedly, it would make sense to proceed step by step and with complete transparency. Thus, instead of providing only adjusted values or weights on SIPP public-use files, it would probably be better to provide the reported values and unadjusted weights with separate fields containing the adjustment factors. Users could then make their own evaluations and decisions as to which set of values to use. For example, researchers modeling behavioral responses to tax and transfer policies may prefer to use reported amounts rather than adjusted amounts because respondents' behavior may be affected more by their belief about the size of a payment than by the actual size of a payment.

Arguing in favor of proceeding down this road of adjustment is the checkered history of work to improve survey response. Census Bureau survey researchers and others have made major efforts since the days of the Income Survey Development Program to develop the best possible questionnaire design and interviewer training to elicit accurate reports of income and program participation from survey respondents. Yet these efforts have met with mixed success. While reporting of many types of income and program participation in SIPP is better than in other surveys, SIPP still exhibits significant net reporting errors for key programs, and the quality of reporting for some programs has declined rather than improved over time. Moreover, SIPP captures only about 80 percent as much aggregate wages as the CPS, and, given that wages are about 78 percent of total household income, the SIPP estimate of total income suffers significantly as a result.

Given users' needs for data that are as accurate as possible and the seeming inability to obtain better reporting through survey instrumentation alone, we encourage the Census Bureau to actively explore the production of SIPP public-use microdata files that include adjustment factors for income sources and program participation to produce agreement with the best independent estimates. A prime target of opportunity could be the use of state records of employment and earnings that are provided to the LEHD Program to adjust reported values and in other ways improve the quality of employment and earnings data in SIPP. Alternatively, the National Directory of New Hires may prove to be a feasible source of such data for use in SIPP.

Replacing Survey Questions

Given the complexity of developing imputation models and adjustment factors as described above, it might seem preferable to simply use administrative records values, when available, for all survey respondents and to drop the particular items from the questionnaire. In fact, when SSA researchers receive matched files of SIPP and SSA records from the Census Bureau for analysis and simulation modeling of their programs, they routinely replace the survey values with administrative records values for Social Security and SSI benefits (personal communication from Bernard Wixon, Office of

Research, Evaluation, and Statistics, Social Security Administration, to the panel, January 8, 2009).

Depending on the legal authority of the custodial agency, however, there could be high hurdles to obtaining permission for a direct use of administrative records to take the place of survey questions. There would also be issues of timeliness, informed consent, and increased disclosure risk, and some records would not be suitable for this use because of conceptual inconsistencies with the desired survey responses.

Disclosure risk would be greater with using administrative records values instead of asking survey questions, even though, functionally, the data are equivalent in that the survey questions are trying to elicit responses that equal the administrative records values for an individual. In practice, as we have seen, survey reports are often erroneous to a greater or lesser extent, which affords added confidentiality protection compared with the actual administrative records values. Moreover, some of the custodial agency's employees have access to individual administrative records, which could enable them to identify particular people in the survey and inadvertently (or advertently) make this known. To respond to these concerns, some kind of probabilistic perturbation of the administrative records values that are used in place of asking survey questions would be required to provide sufficient confidentiality protection, in addition to agreements with the custodial agency that included penalties for a breach of confidentiality by that agency's employees.

The risks from direct replacement of survey values arise when the substituted administrative records values make the record unusual on quasi-identifiers.[12] In general, this is more likely to occur when substituting several items per record rather than one item. For example, substituting program participation status and not also benefit amounts is less risky than substituting both items, and substituting benefit amounts for 1 month is less risky than substituting benefit amounts for 12 or more months because many records have the same status in any 1 month but fewer have the same annual or multiyear history. The Census Bureau can gauge the severity of disclosure risks from intruders who do not have access to the custodial agency's records by performing experiments that attempt to link SIPP records containing more or fewer substituted items from administrative records with externally available sources. The bureau could also determine the risks from intruders who do have access to the agency records.

[12]Quasi-identifiers—as distinct from name, Social Security number, and similar unique identifiers—are combinations of variables, such as gender, birth date, and zip code, which can make it possible to identify all or most individuals in a data set through matching against external sources.

Nonetheless, the use of administrative records to replace survey values for one or more variables in a SIPP panel, when feasible, would have the benefits of reducing respondent burden and improving data quality. We note, in this regard, that Title 13, Section 6, of the U.S. Code, which pertains to the Census Bureau, authorizes the secretary of commerce as follows:

a. The Secretary, whenever he considers it advisable, may call upon any other department, agency, or establishment of the Federal Government, or of the government of the District of Columbia, for information pertinent to the work provided for in this title.
b. The Secretary may acquire, by purchase or otherwise, from States, counties, cities, or other units of government, or their instrumentalities, or from private persons and agencies, such copies of records, reports, and other material as may be required for the efficient and economical conduct of the censuses and surveys provided for in this title.
c. To the maximum extent possible and consistent with the kind, timeliness, quality and scope of the statistics required, the Secretary *shall* acquire and use information available from any source referred to in subsection (a) or (b) of this section instead of conducting direct inquiries [emphasis added].

As an example of the benefits from direct substitution of administrative records values for survey questions, consider Social Security and SSI benefits. They are among the best reported income sources in SIPP (and other surveys), with 90-91 percent of aggregate benefits typically reported and even higher percentages of participation reported in the aggregate (Meyer, Mok, and Sullivan, 2009). Yet Huynh, Rupp, and Sears (2001), summarized above, identified individual reporting errors for these programs based on a matched SSA-SIPP file. Moreover, Social Security benefits are such an important component of income for the elderly population that adding even as little as 8-10 percent more benefit dollars to SIPP through replacing survey reports with values from SSA records could make a significant difference in the poverty status for this group.

Adding New Variables

A third direct use of administrative records is to add variables to a survey that are not and have never been included in the questionnaire but that could be useful to append for policy analysis and research purposes. The SIPP gold standard project described above is an example. This project involved augmenting SIPP data files with exactly matched administrative records on earnings histories and Social Security benefits. In addition, because the gold standard file can be used only at the Census Bureau, the project is intended to find a way, through state-of-the-art synthesizing techniques, to deliver a useful public-use microdata file for retirement policy analysis that contains the linked survey reports, longitudinal earnings records, and Social Security benefit records. (See "Confidentiality Protection and Data Access" below for a discussion of synthesizing techniques and alternative modes of data access.)

This work has involved dedicated effort and leading-edge thinking by Census Bureau staff and academic researchers, but the results to date are mixed. Early, limited analysis by Abowd (2008) found that the synthesized public-use version of the gold standard file adequately represented the patterns of earnings histories in the data for some demographic groups but not others and underestimated early retirement and retirement at age 65. The more detailed evaluation commissioned by SSA (Urban Institute/NORC Evaluation Team, 2009) found that many univariate distributions were accurately represented in the synthetic file, but that the results for regression analyses and policy simulations were more mixed. There were many differences in simulation results that would have led researchers to erroneous conclusions by using the synthetic file. Another important problem in the synthesized file was an overestimation of the duration of marriage, which has implications for analysis of retirement and income security. Other problems found in the synthetic file were present in the gold standard file itself and not produced by the synthesization. Overall, the evaluation team concluded (pp. 1-5) that "the effort to synthesize on a such a large scale was a 'bridge too far,' given how early the whole profession is in creating and using synthetic data" but that the work is promising, particularly if undertaken on a smaller scale.

In general, synthetic public-use data present the problem that the synthesized data are not likely to preserve relationships among variables that are not the focus of the synthesizing effort—for example, the relationship of immigrant parental income and children's educational attainment in SIPP. Furthermore, like all statistical models, synthesis models are approximations of reality, so that they may not accurately capture some distributional features in the original data. Consequently, some SIPP data users may not find a fully, or almost fully synthesized public-use file, such as the gold standard file, useful for their needs and, to work with the actual linked data, would have to go to an RDC. Some users are averse to RDCs, which can involve extensive and long approval processes (see National Research Council, 2005).

We encourage the Census Bureau to consider carefully the benefits and costs of appending administrative records data to SIPP files for public use. When new variables are appended, particularly detailed longitudinal histories, such as longitudinal earnings records, the increase in disclosure risk is likely to be substantial, even when an intruder does not have access to the custodial agency's records. Alternative approaches are possible, however. One approach is to transform the appended data into categorical instead of continuous variables. In the case of earnings histories, for example, categorical variables could represent different patterns of earnings histories (number of lifetime jobs, number of periods out of the labor force, etc.) rather than the detailed histories. Another approach (which could be used in combination with categorization of selected variables) is to use partial synthesis of a much smaller set of selected values. Such partial synthesization could provide reliable information with satisfactory confidentiality protection, as we discuss below. In any event, the need for appending additional variables to SIPP should be carefully vetted with data users because of the implications for confidentiality protection and data access.

CONFIDENTIALITY PROTECTION AND DATA ACCESS

As summarized in Box 3-1, the Census Bureau, like other data disseminators that collect individual information under a pledge of confidentiality, strives to release data files that are not only safe from illicit efforts to obtain respondents' identities or sensitive attributes, but also useful for analysis. In general, strategies for optimizing the risk-utility trade-off fit into two broad categories. *Restricted access strategies* allow only select analysts to use the data, for example, via licensing or by requiring analysts to work in secure data enclaves. *Restricted data strategies* allow analysts to use altered versions of the data, for example, by deleting variables from the file, aggregating categories, or perturbing data values (see National Research Council, 2005). The Census Bureau has extensive experience in applying both of these methods. For example, currently, standard public-use files of SIPP data (not linked with administrative records) can be downloaded from the SIPP website, and a version of SIPP data for specific panels linked with earnings histories and Social Security benefits can be used in the RDCs (the gold standard project). Both restricted data and restricted access strategies are likely to be useful for a reengineered SIPP, as described below.

Restricted Data for SIPP

The Census Bureau releases public-use microdata samples for many of its products, including SIPP, usually with some values altered to protect confidentiality. Typical alterations include

- recoding variables, such as releasing ages or geographical variables in aggregated categories;

- reporting exact values only above or below certain thresholds, for example, reporting all incomes above $100,000 as "$100,000 or more";
- swapping data values for selected records, for example, switching the quasi-identifiers for at-risk records with those for other records to discourage users from matching, since matches may be based on incorrect data; and
- adding noise to numerical data values to reduce the possibilities of exact matching on key variables or to distort the values of sensitive variables.

These methods can be applied with varying intensities. Generally, increasing the amount of alteration decreases the risks of disclosures; but, it also decreases the accuracy of inferences obtained from the released data, since these methods distort relationships among the variables. For example, aggregation makes analyses at finer levels impossible and can create ecological inference problems, and intensive data swapping severely attenuates correlations between the swapped and unswapped variables. It is difficult—and for some analyses impossible—for data users to determine how much their particular estimation has been compromised by the data alteration, in part because disseminators rarely release detailed information about the disclosure limitation strategy. Even when such information is available, adjusting for the data alteration to obtain valid inferences may be beyond some users' statistical knowledge. For example, to analyze properly data that include additive random noise, users should apply measurement error models (Fuller, 1993) or the likelihood-based approach of Little (1993), which are difficult to use for nonstandard estimands.[13] Nonetheless, when the amount of alteration is very small, the negative impacts of traditional disclosure limitation methods on data utility could be minor compared with the overall error in the data caused by nonresponse and measurement errors.

The current SIPP public-use files (without linked administrative records values) are protected mainly by top-coding monetary variables and age and by suppressing geographic detail in areas with fewer than 250,000 people. In addition, some individuals in metropolitan areas are recoded to be in nonmetropolitan areas with too few people in the sample. This can invalidate estimates of characteristics in nonmetropolitan areas.

---

[13]Estimands are types of estimates, such as means, ranges, percentiles, and regression coefficients.

Protecting Files with Linked SIPP and Administrative Records Data

If values available in administrative data are included in SIPP public-use files, top-coding and geographic aggregation may not offer sufficient protection. The Census Bureau probably would need to alter the administrative variables to prevent exact linking, especially if multiple variables for the same person are culled from an administrative database to create a SIPP record. Additional aggregation, such as rounding monetary values, may offer sufficient protection without impairing data utility. Alteration with high intensity, however, such as intense swapping or noise addition, will attenuate relationships and distort distributions so that the released data are no longer useful.

If heavy substitution of administrative values is planned, one option is to create multiply imputed, partially synthetic data. These data comprise the units originally surveyed with only some collected values replaced with multiple imputations. For example, the Census Bureau could simulate sensitive variables or quasi-identifiers for individuals in the sample with rare combinations of quasi-identifiers, and it might synthesize those values that are available and potentially linkable in external databases.

Partial Synthesis

To illustrate how partially synthetic data might work in practice, we modify the setting described by Reiter (2004). Suppose a statistical agency has collected data on a random sample of 10,000 people. The data comprise each person's race, gender, income, and years of education. Suppose the agency wants to replace race and gender for all people in the sample—or possibly just for a subset, such as all people whose income is below $5,000—to disguise their identities. The agency could generate values of race and gender for these people by randomly simulating values from the joint distribution of race and gender, conditional on their education and income values. These distributions would be estimated using the collected data and possibly other relevant information. The result would be a partially synthetic data set. The agency would repeat this process, say, 10 times, and these 10 data sets would be released to the public.

The analyst would estimate parameters and their variances in each of the synthetic data sets and combine the results using the methods of Reiter (2003). Several statisticians in the Statistical Research Division of the Census Bureau and in academia are working to develop partially synthetic, public-use data for Census Bureau products. These products include the Longitudinal Business Database, the Longitudinal Employer-Household Dynamics data sets, the ACS group quarters, veterans, and full sample data, and the SIPP linked with Social Security benefit information.

Partially synthetic data sets can have positive features for data utility. When the synthetic data are simulated from distributions that reflect the distributions of the collected data, valid inferences for frequencies can be obtained for wide classes of estimands (e.g., means, ranges, percentile distributions). This is true even for high fractions of replacement, whereas swapping high percentages of values or adding noise with large variance produces worthless data. The inferences are determined by combining standard likelihood-based or survey-weighted estimates; the analyst need not learn new statistical methods or software to adjust for the effects of the disclosure limitation. The released data can include simulated values in the tails of distributions so that no top-coding is needed. Finally, because many quasi-identifiers can be simulated, finer details of geography can be released, facilitating small-area estimation.

There is a cost to these benefits—the validity of synthetic data inferences depends on the validity of the models used to generate the synthetic data. The extent of this dependence is driven by the nature of the synthesis and the question asked. For example, when all of race and gender are synthesized, analyses involving those variables would reflect only the relationships included in the data generation models. When the models fail to reflect certain relationships accurately, analysts' inferences also would not reflect those relationships. Similarly, incorrect distributional assumptions built into the models would be passed on to the users' analyses. However, when replacing only a select fraction of race and gender and leaving many original values on the file, inferences may be relatively insensitive to the assumptions of the synthetic data models.

In practice, this model dependence means that agencies should release metadata that help analysts decide whether or not the synthetic data are reliable for their analyses. For example, agencies might include the code used to generate the synthetic values as attachments to public releases of data. Or they might include generic statements that describe the imputation models, such as "main effects and interactions for income, education, and gender are included in the imputation models for race." Analysts who desire finer detail than afforded by the imputations may have to apply for restricted access to the collected data.

Even with such metadata, secondary data analysts would be greatly helped if the Census Bureau provided some way for them to learn in real time about the quality of inferences based on the synthetic data (or any masked version of SIPP). Ideally, the quality measures provided would be specific to particular inferential quantities rather than broad measures. For example, reporting comparisons of means, variances, and correlations in the observed and synthetic data does little to help analysts estimating complex models.

One approach is for the Census Bureau to develop a verification server

(Reiter, Oganian, and Karr, 2009). This server, located at the Census Bureau, would store the original and synthetic (or otherwise masked) data sets. Analysts, who have only the synthetic data, would submit queries to the server for measures of data quality for certain estimands. The server would run the analysis on both the original and synthetic data and report back to the analyst a measure of data quality that compares the inferences obtained from both sources. The server could also serve as a feedback mechanism for the agency, capturing what quantities analysts care most about. Agencies might be able to use this information to improve the quality of future data releases. There may be additional disclosure risks of releasing the utility measures; research would be needed to gauge these risks and, more broadly, to develop and fully test the functionality and usability of a verification server.

Synthesizing SIPP Data

The synthesis of the SIPP gold standard file, which contains linked SIPP, SSA, and IRS data, is very intense: Only a handful of some 600 variables remain unsynthesized. Practically all variables are synthesized to ensure a small chance of linking the synthesized records to the existing SIPP public-use records. With the reengineered SIPP, such heavy synthesis may not be necessary. If the released data do not include such detailed administrative information as longitudinal earnings histories, the Census Bureau can synthesize only the values of quasi-identifiers for at-risk records and the linkable values available in administrative sources. It may not even be necessary to synthesize entire variables to achieve adequate protection. For example, synthetic values could replace top-coded monetary and age values and aggregated geographies. The benefits of synthesis over top-coding are illustrated by An and Little (2007); more research is needed on methods for simulating geographies. Providing information in the tails and finer geographies would improve on the current SIPP public-use product without necessarily increasing disclosure risks. Methods of gauging the risks inherent in partially synthetic data with only some values synthesized are described in Reiter and Mitra (2009).

If the released data do contain detailed administrative data, similar to the gold standard file, the Census Bureau has several options. It can proceed as with the current SIPP, releasing a file without linked data and a highly synthesized version of the linked data. Or it can try to reach new memoranda of understanding with SSA and IRS that make it possible to do less synthesizing. For example, it may be possible to synthesize earnings and benefits histories, leaving the other variables on SIPP as is. Regardless of the path chosen, the Census Bureau should recognize that most SIPP users are not likely to support the release of a file with linked administrative records if the time required to create the file and evaluate its risks and utility delays its release in comparison to a standard SIPP public-use file.

Restricted Access for SIPP

In addition to public-use microdata files, the Census Bureau makes more detailed data from SIPP and other surveys available via a restricted access mode, which permits use of the data in any of the nine RDCs operated by the bureau (see http://www.ces.census.gov/index.php/ces/cmshome). The files available in the RDCs are stripped of obvious identifiers, such as name and address, but do not contain recodes or other modifications that blur the underlying data in the public-use versions.[14]

The RDC restricted access mode, however, has limitations. Analysts who do not live near a secure data enclave, or who do not have the resources to relocate temporarily to be near one, are shut out from RDCs. Gaining restricted access generally requires months of proposal preparation and background checks; analysts cannot simply walk into any secure data enclave and immediately start working with the data. As recommended by a previous National Research Council report (2005), the Census Bureau should continue to pursue ways to speed up the project approval process in the RDCs.

Another restricted access approach is to establish a remote access system for SIPP data. When queried by analysts, these systems provide output from statistical models without revealing the data that generated the output. Such servers are in the testing stage at the Census Bureau. If they are found useful, they would provide an excellent resource for certain analyses on the genuine data without having to go to an RDC. However, remote access systems are not immune from disclosure risks. Clever queries can reveal individual data values. For example, asking for a regression model that includes an indicator variable that equals 1 for a unique value of some predictor and 0 for all other variables enables the analyst to predict the outcome variable perfectly (Gomatam et al., 2005). These types of intrusions could be especially problematic if a public-use data set is provided and the remote access system is open to all users. For example, an ill-intentioned user could look at a continuous, unaltered variable to determine unique values, then submit regression queries with indicator variables to learn about those records' other variables. The Census Bureau can limit the risks of such problems by restricting access to the server. For example, users of the server could be required to go through a licensing procedure. In addition, the server could keep track of and audit requests,

[14]To date, SIPP files that have been linked to administrative records are not available in the RDCs outside the Census Bureau.

so that any ill-intentioned intruder who sneaks through the licensing might be identified and punished.

CONCLUSIONS AND RECOMMENDATIONS

The Role of Administrative Records in a Reengineered SIPP

Conclusion 3-1: In reengineering the Survey of Income and Program Participation (SIPP) to provide policy-relevant information on the short-run dynamics of economic well-being for families and households, the Census Bureau must continue to use survey interviews as the primary data collection vehicle. Administrative records from federal and state agencies cannot replace SIPP, primarily because they do not provide information on people who are eligible for—but do not participate in—government assistance programs and, more generally, because they do not provide all of the detail that is needed for SIPP to serve its primary goal. Many records are also difficult to acquire and use because of legal restrictions on data sharing, and some of the information they contain may be erroneous. Nonetheless, information from administrative records that is relevant to SIPP and likely to improve the quality of SIPP reports of program participation and income receipt in particular can and should be used in a reengineered SIPP.

Conclusion 3-2: The Census Bureau has made excellent progress with the Statistical Administrative Records System and related systems, such as the person validation system, in building the infrastructure to support widespread use of administrative records in its household survey programs. The bureau's administrative records program, both now and in the future as it adds new sets of records and analysis capabilities, will be an important resource for applications of administrative records in a reengineered Survey of Income and Program Participation.

Acquisition of Records

Conclusion 3-3: Many relevant federal administrative records are readily available to the Census Bureau for use in a reengineered Survey of Income and Program Participation (SIPP). However, most state administrative data are not available for use in a reengineered SIPP at this time and could be difficult to obtain.

Recommendation 3-1: The Census Bureau should seek to acquire additional federal records that are relevant to the Survey of Income and Program Participation, which could include records from the U.S. Department of Veterans Affairs and the Office of Child Support Enforcement.

Recommendation 3-2: The Census Bureau, in close consultation with users, should develop a strategy for acquiring selected state administrative records, recognizing that it will be costly and probably unfeasible to acquire all relevant records from all or even most states. The bureau's acquisition strategy should be guided by such criteria as the importance of the income source for lower income households, particularly in times of economic distress, and the relative ease of acquiring the records. Unemployment insurance benefit records should be a high priority for the Census Bureau to acquire on both of these counts, and the bureau should investigate whether it is possible to acquire these records from the National Directory of New Hires, which would eliminate the need to negotiate with individual states.

Indirect Uses of Records

Conclusion 3-4: Indirect uses of administrative records are those uses, such as evaluation of data quality and improvement of imputation models for missing data, in which the administrative data are never recorded on survey records. They are advantageous for a reengineered Survey of Income and Program Participation (SIPP) in that they should have little or no adverse effects on timeliness or the needed level of confidentiality protection of SIPP data products.

Recommendation 3-3: The Census Bureau, in close cooperation with knowledgeable staff from program agencies, should conduct regular, frequent assessments of Survey of Income and Program Participation (SIPP) data quality by comparison with aggregate counts of recipients and income and benefit amounts from appropriate administrative records. When feasible, the bureau should also evaluate reporting errors for income sources—both underreporting and overreporting—by exact-match studies that link SIPP records with the corresponding administrative records. The Census Bureau should use the results of aggregate and individual-level comparisons to identify priority areas for improving SIPP data quality.

Recommendation 3-4: The Census Bureau should move to replace hot-deck imputation routines for missing data in the Survey of Income and Program Participation with modern model-based imputations, implemented multiple times to permit estimating the variability due to imputation. Imputation models for program participation and benefits should make use of program eligibility criteria and characteristics of beneficiaries from administrative records so that the imputed values reflect as closely as possible what is known about the beneficiary population. Before implementation, new imputation models should be evaluated to establish their superiority to the imputation routines they are to replace.

Recommendation 3-5: The Census Bureau should request the Statistical and Science Policy Office in the U.S. Office of Management and Budget to establish an interagency working group on uses of administrative records in the Survey of Income and Program Participation (SIPP).[15] The group would include technical staff from relevant agencies who have deep knowledge of assistance programs and income sources along with Census Bureau SIPP staff. The group would facilitate regular comparisons of SIPP data with administrative records counts of income recipients and amounts (see Recommendation 3-3) and advise the Census Bureau on priorities for acquiring additional federal and selected state administrative records, how best to tailor imputation models for different sources of income and program benefits, and other matters related to the most effective ways to use administrative records in SIPP. The Census Bureau should regularly report on its progress in implementing priority actions identified by the group.

Direct Uses of Records

Conclusion 3-5: Direct uses of administrative records in a reengineered Survey of Income and Program Participation (SIPP), which include substituting administrative values for missing survey responses, adjusting survey responses for net underreporting, using administrative values instead of asking survey questions, and appending additional administrative data, potentially offer significant improvements in the quality of SIPP data on income and program participation. They also raise significant concerns about increased risks of disclosure and delays in the release of SIPP data products.

Recommendation 3-6: In the near term, the Census Bureau should give priority to indirect uses of administrative records in a reengineered Survey of Income and Program Participation (SIPP). At the same time and working closely with data users and agencies with custody of relevant administrative records, the bureau should identify feasible direct uses of administrative records in SIPP to be implemented in the medium and longer terms. Social Security and Supplemental Security Income benefit records, which are available to the Census Bureau on a timely basis, are prime candidates for research and development on ways to use the administrative values directly—either to adjust survey responses for categories of beneficiaries or to replace survey questions (which would reduce respondent burden)—in ways that protect confidentiality.

---

[15]See Recommendation 4-5 regarding an advisory group of outside researchers and policy analysts.

Recommendation 3-7: When considering the addition to the Survey of Income and Program Participation (SIPP) of administrative records values for variables that have never been ascertained in the survey itself, the Census Bureau should ensure that the benefits from the added variables are worth the costs, such as additional steps to protect confidentiality. The bureau should consult closely with users to be sure that the added variables are central to SIPP's purpose to provide information on the short-run dynamics of economic well-being and that their inclusion does not compromise the ability to release public-use microdata files that accurately represent the survey data.

Confidentiality Protection and Data Access

Conclusion 3-6: Multiple strategies for confidentiality protection and data access are necessary for a survey as rich in data as the Survey of Income and Program Participation. Public-use microdata files, which are available on a timely basis and in which confidentiality protection techniques do not unduly distort the relationships in the data, are the preferred mode of data release. Some uses may require access to confidential data that at present can be provided only at one of the Census Bureau's Research Data Centers.

Recommendation 3-8: The Census Bureau should develop confidentiality protection techniques and restricted access modes for the Survey of Income and Program Participation (SIPP) that are as user-friendly as possible, consistent with the bureau's duty to minimize disclosure risk. In this regard, the bureau should develop partial synthesis techniques for SIPP public-use microdata files that, based on evaluation results, are found to preserve the research utility of the information. For SIPP data that cannot be publicly released, the Census Bureau should give high priority to developing a secure remote access system that does not require visiting a Research Data Center to use the information. The bureau should also deposit SIPP files of linked survey and administrative records data (with identifiers removed) at all Research Data Centers in order to expand the opportunities for research that contributes to scientific knowledge and informed public policy.

4

Innovation in Design and Data Collection

The Census Bureau's plans for the redesigned Survey of Income and Program Participation (SIPP) have three primary elements. The first, to make greater use of administrative data to improve data quality, is discussed in Chapter 3. The second, to improve the processing system for SIPP, involves converting a computer-assisted personal interview (CAPI) survey instrument that is currently implemented in an obsolete survey questionnaire programming language to the widely used Windows-based BLAISE survey programming language (see http://www.blaise.com/?q=ShortIntroduction). Moreover, the Census Bureau is converting the postinterview data processing system from Fortran to SAS and is improving the documentation of SIPP data editing and imputation procedures. The panel commends the Census Bureau's efforts in these important undertakings. The panel has the general belief that these are worthwhile, constructive steps, but they were outside the scope of the panel's review. Hence, the panel says nothing further about them.

The third element is to change SIPP from its current structure, in which interviews are conducted every 4 months for each of four staggered rotation groups (thus ensuring a uniform month-by-month workload for SIPP interviewers), to an annual interview making use of an event history calendar (EHC) to document intrayear changes in demographic and economic circumstances. Regularly scheduled topical modules will no longer be included in the redesigned SIPP, although some prior topical module content will be incorporated into the primary survey instrument, and federal agencies may pay for supplemental questions to be asked between annual interviews.

SIPP sample members will be followed for 3 to 4 years, but, following SIPP practice since 1996, the panels will not overlap.

The first part of this chapter discusses concerns about moving SIPP to a nonoverlapping annual survey that relies on EHCs to develop month-to-month information on households. The remainder of the chapter discusses several additional issues related to SIPP design features (length and frequency of interviews, length and overlap of panels), content, timeliness, and budget that the panel thinks are important.

One feature of SIPP that the panel does not discuss is the sample size and design. The current design (see Chapter 2), which oversamples low-income populations based on the previous census, has been in use beginning with the 1996 panel, and sample sizes have been what the SIPP budget could afford. While data users would always prefer additional sample, SIPP users have found the sample sizes of recent SIPP panels (see Table 2-1) to be adequate for most purposes. The design, although not state representative, includes cases in every state (most of which are identified on the public-use microdata files) so that researchers can take account of differences in state tax and transfer program rules in their analyses. Ordinarily, the design would next be revised based on the 2010 census; however, that census will not include a long-form sample with data on income and other socioeconomic characteristics. Instead, the continuous American Community Survey (ACS) now provides that information (beginning in 2005). It will be necessary to redesign the SIPP sample to use the ACS, but it is our understanding that the ACS will not be available until 2012 for this purpose. As the ACS is relatively new and the shape of the reengineered SIPP is not finalized, the panel thinks it would be premature to comment on sample design issues.

EVENT HISTORY CALENDARS

As emphasized throughout this report, a unique feature of SIPP is its capacity to measure short-run dynamics. Monthly data on incomes, employment, program participation, health insurance coverage, and demographic characteristics of the household allow analysts to study transitions into marriage and divorce, transitions into and out of poverty, and transitions in health insurance coverage, at a monthly frequency. Monthly data also make SIPP particularly well suited for assessing eligibility for major transfer programs, since program rules typically depend on economic and demographic characteristics in the month or months prior to application. Studies of program take-up require careful calculations of eligibility—the denominator of the take-up rate—and high-quality measures of program participation—the numerator of the take-up rate. Studies of short-run dynamics are impossible with other nationally representative data sets, and studies of take-up are badly flawed if the period reflected in the data does not align with the period over which program eligibility is assessed. In short, the monthly time frame is essential for many of the applications that use SIPP data. The Census Bureau's plans to move SIPP to an annual survey, filling in intrayear dynamics using EHCs, potentially affects—perhaps positively, perhaps negatively—SIPP's single most important feature.

What Is an Event History Calendar?

An EHC interview is centered on a customized calendar that shows the reference period under investigation (1 year in the case of the reengineered SIPP). The calendar contains time lines for different domains, for example, residence history, household composition, work history, and other areas that might be helpful in aiding the respondent's memory. As discussed in Belli (1998), in an EHC, "respondents are encouraged to consider various events that constitute their personal pasts as contained within broader thematic streams of events. Not only can respondents note the interrelationship of events within the same themes (top-down and sequential retrieval) but, depending on which themes are represented by the calendar, respondents can also note the interrelationships among events that exist within different themes (parallel retrieval)." Put more concretely, if respondents tend to remember life events as "I lost my job a month after having my second baby," interview accuracy may improve if respondents are allowed to connect these events in calendar time, rather than reporting births in a household roster and job changes later in the interview in an employment section of the questionnaire.

Another potential advantage of the EHC approach, if it proves capable of generating high-quality monthly data, is that the first year of income data could be collected with no added sample attrition beyond the loss of households that refuse to participate in the survey at all. This is a substantial potential advantage relative to the data conventionally collected in SIPP. Under the current design, annual income must be aggregated across four waves in order to have a common 12-month reference period for the four rotation groups. Annual income for the first calendar year of the conventionally collected SIPP panel requires data through Wave 4, which will be affected by three waves of attrition beyond the initial sample loss at Wave 1. In the 2004 SIPP panel, the cumulative sample loss after four waves was 28 percent compared with a Wave 1 nonresponse rate of 15 percent (from information provided by Census Bureau staff; see also Table 2-1 in Chapter 2).

Several ongoing surveys make use of EHCs for at least a portion of their survey content, including the Panel Study of Income Dynamics (PSID), the 1997 National Longitudinal Survey of Youth (NLSY97), the

Los Angeles Family and Neighborhood Survey, and the British Panel Survey. In December 2007, leaders at the Census Bureau and the Panel Study of Income Dynamics convened a conference of survey design experts and other scholars knowledgeable about event history methodology to learn from and improve their plans (see http://psidonline.isr.umich.edu/Publications/Workshops/ehc-07papers.html). The panel commends the Census Bureau for sponsoring this conference and reaching out to additional experts in this methodology.

The Census Bureau and Panel Study of Income Dynamics conference highlighted many of the reasons the Census Bureau is envisioning that an event history methodology may play a key role in the reengineered SIPP's efforts to reduce burden on respondents, reduce program costs, improve accuracy, and improve timeliness and accessibility. Belli (2007) noted that EHCs are "expected to provide advantages to data quality by encouraging respondents to use idiosyncratic cues available in the chronological and thematic structures of autobiographical memory." Fields and Moore (2007) noted that the approach may mitigate missing or erroneous responses by developing timelines for major life events. In particular, the EHC can gather information on overlapping events (such as multiple transfer program participation) or nonoverlapping events (such as a succession of jobs). Moreover, the status at the end of a previously reported calendar year could, in principle, be preloaded to help control seam problems (subject to the respondent being able to override the prior response). If a single annual EHC interview could replace three conventional interviews gathering retrospective information from the prior 4 months, the cost savings could be significant.

There is considerable evidence that the event history methodology can be used successfully to identify demographic changes to a household—the arrival and departure of children, spouses, and other family members—and to identify employment transitions. Both types of events are generally regarded as major life transitions, and it is perhaps not surprising that calendar time may be a convenient way to elicit accurate recall of major life transitions. It is less clear, however, that recall over a 12-month period will be similarly precise for potentially less consequential life events, such as whether a household received benefits from the Special Supplemental Nutrition Program for Women, Infants and Children (WIC) 11 months earlier or when a pay raise occurred for a household member.

The panel is not aware of conclusive evidence that a 12-month EHC framework is capable (or not) of generating accurate information on program participation and income. The Census Bureau recently presented preliminary results from a 2006 paper test of the EHC approach, discussed below, in which it claimed success for the EHC. However, these results (also discussed below) were limited in scope and showed a mixed picture with regard to the ability of the EHC to accurately capture monthly income.

Several passages in the papers prepared for the Census Bureau's EHC conference highlighted the uncertainty associated with the approach. Sastry, Pebley, and Peterson (2007:20), writing about the Los Angeles Family and Neighborhood Survey, conclude, "we recommend keeping the period covered by the EHC to a minimum and only using it to collect information on domains and topics that are difficult to collect using standard question-list approaches." Callegaro and Belli (2007) suggest that the EHC approach may reduce seam bias, but they also expect that the magnitude of the seam effect will increase when moving from quarterly to yearly data collection. In a different paper, Belli (2007:13), writing about an experimental subsample of the PSID, finds "with program participation, the [conventional questionnaire] showed consistent advantages in reports among disadvantaged groups in comparison to the event history calendar for the timing of receipt of benefits during 1996." Pierret and colleagues (2007:28), writing about the NLSY97, note: "one decision that we have made is not to collect all details on every spell for every event history. This decision reflects our experience that many respondents have difficulty recalling details of events that occurred far in the past and lasted a very short time." This conclusion is troubling for the proposed changes to SIPP, since the interview time frame for the NLSY97, like the reengineered SIPP, is 1 year.

Testing the EHC Approach for SIPP

The lack of evidence about the ability of an EHC to collect monthly data on the many topics that are covered in SIPP places considerable pressure on the Census Bureau. Not only must the bureau design an effective pretesting program for the EHC methodology, but it must also make its survey reengineering plans for SIPP sufficiently flexible so that it can modify its plans if the pretesting reveals unanticipated, negative evidence on the likely success of the proposed methodology.

Paper EHC Test

The Census Bureau administered a paper test of the EHC approach that was completed in June 2008. This test was designed primarily to give the bureau a relatively quick "go/no-go" signal for continued investment in further development of an automated instrument and larger scale testing. The sample for this test was drawn from 2004 SIPP panel participants from Illinois and Texas. A point of emphasis in the paper test was on designing and administering the EHC instrument. Given this, professionals from the Census Bureau and the Office of Management and Budget observed a large number of paper test interviews. Assessments from observation reports and field representative debriefing reports, in addition to comparisons of estimates from the standard SIPP and EHC questionnaires and comparisons with administrative records for selected programs, will be obtained, with the goal of furthering knowledge about the overarching question: Can the EHC methodology produce data of similar quality to that of the standard SIPP interview?

The Census Bureau recently presented preliminary findings from the 2008 paper test based on comparing aggregate reports of selected income sources and other characteristics from the standard SIPP questionnaire and the EHC questionnaire for 1,620 cases that completed both types of questionnaires (Moore et al., 2009). The results are both promising and disquieting. For SSI and WIC (Illinois only), the aggregate estimates of recipients track very closely for the months of January-December 2007. For Medicare, Social Security, WIC (Texas only), and food stamps (Illinois only), aggregate estimates of recipients show the same patterns over the 12-month period, but the EHC levels are significantly lower than the standard questionnaire levels—by several percentage points for Medicare, for example. For food stamps (Texas only), employment, and school enrollment, the trends in monthly aggregates differ between the standard and EHC questionnaires—for example, the standard questionnaire aggregates are several percentage points higher than the EHC aggregates in January-September 2007 and about the same as the EHC aggregates in the rest of the year. No results have been presented as yet on comparisons of benefit amounts, on the extent to which the standard and EHC responses track across time on an individual respondent basis, or on comparisons with administrative records, which will involve the entire test sample, including SIPP participants who were cut from the sample in 2006 and so did not respond to the standard SIPP questionnaire for 2007.

The panel commends the Census Bureau for conducting this paper test of the EHC methodology. It undoubtedly will provide valuable information on ways to administer the calendars in the context of a comprehensive national survey. Moreover, it will provide the first available information on the ability of households to recall spells of program participation and amounts of monthly income. Nevertheless, an extensive program of design and research must be conducted to assess the EHC approach. We describe a set of unresolved issues below.

First, more needs to be learned about how data collection mode affects content. The paper test, of course, uses a different mode than the BLAISE-based computer-assisted interviewing that is envisioned for the reengineered SIPP. There is evidence in some contexts that survey mode (e.g., paper versus computer) has relatively minor effects on survey responses in some domains (see Carini et al., 2003), but that respondents tend to prefer computer-based applications. If so, particularly for a long, time-intensive survey like SIPP, the paper test may understate the ability of the EHC approach to elicit accurate information if people are put off by the structure of the paper test. Alternatively, the specially trained interviewers for the paper test may aid respondents in a manner that would not occur for the reengineered SIPP. At a minimum, the discrepancy between the paper test and the actual data collection mode that will be used raises one concern about the value of the paper test results.

Second, samples used for a test of the EHC approach need to be large enough to generate reliable results. To give a sense of the sampling difficulties that EHC tests face, consider the following: in 2006, about 8.9 percent of the U.S. population received food stamp benefits, whereas only about 2.4 percent received Supplemental Security Income (SSI) benefits and only about 1.6 percent received Temporary Assistance for Needy Families (TANF) benefits (Assistant Secretary for Planning and Evaluation, 2008:Tables IND 3a, 3b, 3c). Given these figures, serious tests of the EHC need large samples to ensure there are a substantial number of respondents receiving TANF benefits, SSI benefits, or food stamps. This can be done by making appropriate power calculations and then drawing appropriately sized test samples, perhaps augmented by oversamples of program recipients drawn from administrative records. If too few program participants are in an EHC test sample, it will be extremely difficult for the Census Bureau to assess whether the EHC can provide accurate month-to-month information on program participation for sampled individuals. The problem is even more acute if the test is to provide useful information on multiple program participation, since even smaller fractions of the population will simultaneously participate in more than one program. Facilitating accurate analysis of program participation is one of the central goals of SIPP.

Tests of the EHC face another sample-related concern. The Census Bureau needs to have some benchmark that it can use to assess the quality of EHC responses. Two possibilities suggest themselves. First, the test results can be matched against administrative data. The Census Bureau is pursuing this approach. The paper test includes matching the survey results to data drawn from administrative records on program receipt in Illinois and Texas. This raises the question mentioned above: Are SIPP samples from Texas and Illinois large enough to provide a reasonable assessment of the EHC approach? In addition, can results for a sample from Texas and Illinois be generalized to the U.S. population? A subsequent "electronic prototype" test, described below, will add more states to the evaluation, which is a positive step forward. The second benchmark would be to field an EHC-based survey concurrently with a traditional SIPP survey, allowing for immediate comparisons of the two approaches. We say more about this possibility below.

A third unresolved issue has to do with the effects of an EHC approach on seam bias and sample attrition. As described in Chapter 2, a major issue for the traditional SIPP is that too many transitions—on and off programs, in and out of the formal labor market, in and out of health insurance coverage—happen at the beginning of a new survey wave. Moreover, large percentages of sample participants leave the survey following the first wave. It is not clear how the EHC approach will affect these problems. By having fewer waves (or seams), seam bias may be diminished. But transitions may pile up at the point of annual sampling, making the longitudinal information elicited from the EHC less valuable. Respondent burdens with the EHC approach are high, since calendars must be used for an extensive set of employment, program, and demographic characteristics. It is not clear how the burdens will affect survey attrition.

Finally, a problem with the 2008 paper test comparisons reported to date is that participants in the "traditional" SIPP were also the sample for the comparisons. These households in the test already provided monthly detail on incomes, employment, demographic changes, insurance coverage, and program participation. This raises the question of whether respondents who have already recorded this information in the SIPP 4-month interviews were better able to respond accurately to the paper EHC than would be the case if the EHC sample cases had all been drawn independently.

Electronic EHC Test

To provide further evidence on these issues, the Census Bureau plans to test a one- or two-wave electronic prototype EHC in early 2010. If funding during FY 2010 and FY 2011 is available, this prototype would examine issues that arise with locating movers when interviews are 1 year rather than 4 months apart, as well as the consistency of data reports between interviews that are 1 year apart. The development and implementation of the prototype experiment is a valuable next step in developing the information base needed for the reengineered SIPP.

The panel does not have enough detail on the 2010 one- or two-wave electronic prototype test to fully assess its ability to resolve questions about whether the EHC approach can adequately replace the traditional SIPP interview structure. Our understanding is that the Census Bureau will not use respondents to the 2008 traditional SIPP panel as the sample for the electronic EHC because of a concern that doing so could compromise responses to the traditional interviews for some or all waves following the 2010 EHC test. Just as important, in our view, is that using a separate sample obviates the concern, expressed above for the paper test, that respondents would provide more accurate reports to the EHC given their participation in the traditional SIPP than if they had not participated in the SIPP.

Instead of using SIPP cases, the Census Bureau plans to conduct EHC interviews in 10 states with about 8,000 households in high-poverty strata that are selected from the areas in which traditional SIPP interviews are currently being conducted. The bureau will then select traditional SIPP cases from the same areas and do side-by-side comparisons of the EHC and SIPP estimates. In addition, the Census Bureau hopes to acquire administrative records from the 10 states that will be used to help evaluate the validity of responses in both the traditional SIPP 2008 panel interviews and the 2010 EHC electronic prototype for calendar year 2009. The panel thinks this broad approach is a promising basis for developing important additional knowledge about the EHC and the traditional SIPP, particularly if the electronic prototype EHC test can be carried out for two waves and not just one wave.

Overlap of Traditional and Reengineered SIPP Panels

While the panel thinks the Census Bureau's EHC electronic prototype plans are promising, it is clear that the knowledge base for EHC methods is not yet sufficiently well developed to have confidence that the approach can be used to generate data of equal or better quality than found in the traditionally collected SIPP. The paper test prototype provides only limited information on data quality for the reasons given above. Moreover, the electronic prototype EHC test, even with its fairly large sample size and even if it is conducted for two waves, is not likely to provide conclusive evidence about the ability of EHCs to match month-to-month details on program eligibility and participation, employment, and income that are obtained with a 4-month interview cycle. Instead, it is likely to provide mixed results, identifying not only strengths but also weaknesses of the EHC approach that require modification and further testing, as well as leaving some issues unresolved—either pro or con.

Consequently, we think it is essential for the Census Bureau to administer (and for Congress to appropriate resources for) a full-blown implementation of the "reengineered" SIPP, concurrently with a traditional SIPP panel. The concurrent surveys should be fielded for at least 2 years, with samples large enough to ensure that a substantial number of survey respondents will in fact be receiving transfer program benefits. Ideally, administrative information on earnings (from the Social Security Administration, SSA), employment (from state employment and wage records), and program participation (from selected state records on TANF and SSA records on SSI and Old-Age, Survivors, and Disability Insurance [OASDI]) would be linked to both surveys, which would allow the Census Bureau to compare aspects of data quality for the traditional and reengineered SIPP designs.

The panel further recommends that the Census Bureau start a new, traditional SIPP panel in February 2012 to provide a comparison data set for the reengineered SIPP panel that will begin in 2013. Respondents who participate over time in longitudinal surveys gain experience in responding to survey questions. Moreover, they are the people who do not leave the survey. Given the experience and selection issues that arise, results from the reengineered SIPP should be compared with the first (rather than fourth) year of a traditional SIPP panel. Assuming the reengineered panel has annual interviews, then the traditional panel with its 4-month interviews must begin a year ahead so that the traditional panel obtains data for the period covered by the first interview of the reengineered panel (2012).

Furthermore, the traditional panel should continue for at least 2 years so that comparisons can be made for at least two interviews of the reengineered panel. Otherwise, it will be impossible to adequately evaluate attrition bias and seam issues that arise in the reengineered SIPP. Moreover, if Wave-1-to-Wave-2 seam bias issues with the reengineered SIPP prove to be a major problem, the Census Bureau can continue to field the traditional SIPP as it further refines the EHC approach. If the expense of having two SIPP surveys in the field is prohibitive, cost savings could be achieved by making the 2012 traditional SIPP panel smaller than prior panels.

There is another reason why it is critical to field overlapping traditional and reengineered SIPP panels. Policy makers, analysts, and researchers who use SIPP need to assess the effects that the new methodology will have on survey findings. One of SIPP's strengths is that it has been fielded since 1984 (with a significant redesign in 1996). Because SIPP panels cover a 25-year period, a common, important use of the data is to document trends in household behavior. As noted earlier, it is clear that problems exist with the traditionally conducted SIPP. But analysts need to have some way of assessing whether changes in trends that arise when comparing results from the reengineered SIPP to results from the traditionally collected SIPP reflect true changes in the population or whether they are a result of changes in survey methodology. The only way to be able to even roughly account for the changes due to survey methodology is to have at least 1 and preferably 2 years of overlapping data.

A third reason to have 2 years of overlap between a traditionally collected SIPP and the reengineered SIPP, in addition to better understanding attrition, seam bias, and the effects of changes in survey methodology, is that responses to the EHC may improve between the first and second interviews. Without a second year of overlap, this improvement would be difficult to detect.

When comparing a full-blown implementation of the reengineered SIPP to a concurrently fielded traditional SIPP or to administrative data, it is important to keep in mind that the form of the measurement error can have important implications for empirical analysis. For example, Gottschalk and Huynh (2006) compare data on inequality from SIPP and detailed administrative earnings records from the Social Security Administration. They show that while SIPP understates inequality, primarily because measurement error is mean-reverting, measures of *mobility* are very similar in SIPP and the administrative data. The point is that all surveys will have error—the importance of error depends on context. Considerable content evaluation has been done with the traditionally collected SIPP over the years. It is critical to have a solid basis for assessing the changes in survey results that arise primarily from changes in survey design, as distinct from changes in respondent behavior. Full overlapping panels are the only way to assess the effects of survey design changes, although they will not necessarily settle all questions about data quality. A third data source, particularly administrative data, would be useful to interpret systematic differences between the reengineered and the traditionally fielded SIPP.

LENGTH AND FREQUENCY OF INTERVIEWS

Respondent Burden Concerns

Moving to an annual schedule of interviews, in which monthly information for an entire year is elicited with EHCs, and continuing to include some topical module content, as planned for the reengineered SIPP, raise concerns that the overall length of the SIPP interview and the burden it places on respondents may exceed that of the current questionnaire. In turn, respondent burden may contribute to item nonresponse, poor quality responses, and attrition from the survey. It is essential, as the Census Bureau evaluates its electronic EHC prototype and implements the overlapping redesigned and traditional SIPP panels, that it not only carefully examine the ability of the EHC approach to generate accurate month-by-month transitions in employment, earnings, household structure, and program participation, but also determine whether the burden on respondents from the redesigned questionnaire is not so taxing as to degrade the overall quality of responses.

The SIPP topical modules have historically provided a large amount of information of considerable interest to the SIPP user community. Many programs have asset tests associated with eligibility rules, making the SIPP asset and liability topical modules essential for accurate modeling of program participation. Other topical modules also contain vital information (see Box 2-1). Yet while the topical modules have provided a great deal of information that is valuable to the fundamental purpose of SIPP, their costs also need to be recognized and weighed against their benefits. Costs include that topical modules require resources that could presumably be used for research and evaluation to improve SIPP; that some topical modules (like the tax topical module) require extensive imputation; and that topical modules may impose burdens on the respondent that could harm the quality of the information gathered by the core questionnaire.

Current redesign plans call for moving some topical module information onto the core SIPP survey, such as the asset and liability module, while other topical module information will be dropped. (Agencies may have the opportunity to gather additional information through reimbursable supplements that would not be fielded at the same time as the core survey.) Undoubtedly decisions will be made that create some controversy in the user community. But as with core content questions in the SIPP redesign efforts, the panel commends the Census Bureau for its exemplary, extensive efforts to solicit information on the needs of the SIPP user community on what topical module information is most important. At the same time, the panel encourages the bureau to measure and take account of respondent burden in making decisions about how much topical module content can be included in the redesigned questionnaire.

Interview length and its consequences for response quality are one factor to consider when thinking about whether the interview periodicity should be 4, 6, or 12 months. For example, some questions, like assets and liabilities, that were in annual topical modules could continue to be asked only once a year, which would allow 6-month or 4-month interviews to be shorter than annual interviews, although the aggregate time spent by respondents in interviews over the year may be longer. The Census Bureau should study the trade-offs in survey quality between longer versus more frequent interviews as part of its research and development program for the reengineered SIPP.

Seam Bias Concerns

The phenomenon of seam bias, in which a large fraction of transitions in employment status, insurance coverage, or program participation occurs between SIPP waves (see Chapter 2), highlights another trade-off that may arise between the traditionally collected and reengineered SIPP. More frequent interviews, as in the traditional SIPP, are widely thought to be helpful in improving the accuracy of survey responses. After all, it is generally easier to remember events that occurred 1 month ago than it is to remember events that occurred 11 months ago. At the same time, frequent interviews also create more opportunities for erroneous reports. Proxy reports, in which one household member provides information for another, are a particular concern that could generate misreporting. A misreported status in one wave creates two false transitions. The trade-off then is between more accurate information that may arise from more frequent interviewing against the higher costs (both financial and, possibly, in the burden on respondents) and the greater likelihood of false transitions that frequent interviewing may induce.

There is inadequate evidence on the causes and empirical importance of false transitions. For example, while SIPP finds a greater share of the population ever without health insurance (over a 12-month period) than does the National Health Interview Survey (which uses a 12-month retrospective question), the differences may simply reflect SIPP's better design for estimating periods without coverage. SIPP also finds more persons ever uninsured than the longitudinal Medical Expenditure Panel Survey (MEPS), which also conducts several interviews during the year. While this might suggest that SIPP does an excellent job of measuring health insurance coverage, the data also show improbable transitions that appear to be reporting errors. Edits to remove improbable transitions—for example, children losing and regaining employer-sponsored coverage through a parent who reported continuous coverage—reduced estimates of children ever uninsured in a year to 24.1 percent from 27.0 percent (Czajka, 2007). There is no publicly available gold standard data source that can be used to benchmark SIPP health insurance coverage transitions. Consequently, more information is needed to assess the trade-off between accuracy on one hand and cost and false transitions on the other.

All of these concerns apply to longitudinal uses of SIPP data, but the impact of annual interviews on cross-sectional estimates needs to be weighed as well. Despite the pronounced seam bias in the present SIPP, the survey's rotation group design and 4-month reference period distribute transitions more or less uniformly across calendar months. This means that only about one-twelfth of all transitions on and off programs occur between December and January, for example. With the proposed annual interviews and a fixed, calendar-year reference period, the Census Bureau runs the risk that three moderately sized seams per year, which are invisible cross-sectionally, will be replaced by one very large seam between December of one year and January of the next.

LENGTH AND OVERLAP OF PANELS

Length

SIPP panels are currently 4 years in length; for the period 1984-1993, they were generally 2-3 years in length. Many panel surveys are much longer in length than SIPP, including the PSID, which has been running since 1968. There are also some panels that are shorter than SIPP, such as the Medical Expenditure Panel Survey, which has panels that collect 2 years of data. There is no way to definitively determine the optimal length for a panel survey; considerations in choosing panel length include the frequency of interviews, the frequency with which there need to be new panels with fresh samples, and the goals and unique contributions of the survey.

Arguing for short panels for SIPP is its focus on providing information on the intrayear dynamics of employment, income, and program eligibility and participation and not on long-term consequences of poverty, welfare dependence, or other phenomena, as in the PSID. Moreover, SIPP's 4-month interview cycle makes it difficult to contemplate a lengthy panel, given the burden on respondents and the resulting panel attrition. SIPP also has a need for new panels at frequent intervals to support its many cross-sectional uses, and the longer each panel runs, the more expensive it becomes to introduce new, overlapping panels (see below). Arguing against very short panels for SIPP, such as 1 or 2 years, is that SIPP users often want to look at changes in income and program participation before and after a major change in program rules or a major event, such as a recession, which requires information on the same respondents over a longer period than just 1 or 2 years.

The current 4-year panel length of SIPP seems about right under the current design of 4-month interviews. Should the change to an annual schedule of interviews using the EHC approach prove successful, then the Census Bureau and the SIPP user community could consider the benefits and costs of lengthening each panel, or perhaps doing so on a periodic basis.

Overlap

Although SIPP is a longitudinal survey, cross-sectional uses of SIPP data abound, such as applications that treat the survey as a source of repeated cross-sections in order to estimate trends. SIPP has monthly cross-sectional weights that are controlled to monthly population totals, and the weights incorporate adjustments for attrition, but these weighting adjustments are not sufficient to make the survey cross-sectionally representative across the full range of characteristics that users might wish to include in their trend analyses or point-in-time estimates (see also Chapter 2). A common strategy for addressing panel bias in a repeated panel survey is to start new panels while earlier panels are still in the field. If the panels overlap in a consistent way, then users can combine panels to produce estimates with uniform bias over time. This strategy was used in SIPP prior to 1996, when new panels were introduced every year, and it is used in other major panel surveys, including MEPS and the Medicare Current Beneficiary Survey. It is also used in the monthly Current Population Survey (CPS), in which the sample consists of eight rotation groups that, for a given month, have been interviewed different numbers of times.

The prior Committee on National Statistics report on SIPP (National Research Council, 1993) noted several advantages of overlapping panels but acknowledged the operational challenges they present for data collection and processing—challenges that prevented SIPP and its users from realizing the full benefits of this design feature in the first decade of the survey's history (National Research Council, 1993). In recognition of both the benefits and challenges of overlapping panels, that report recommended that new panels be started every 2 years, with each panel running for 4 years, instead of 2-3 years. Ideally, this would achieve most of the benefits of an overlapping design but limit to two the number of separate panels fielded and processed at the same time. By comparison, it was often the case in the period 1984-1993 that three panels were in the field at the same time. However, the Census Bureau opted for a 4-year panel in 1996 with no overlap, which permitted a doubling of the panel's size.[1]

The problem of diminishing cross-sectional representativeness over time is likely to persist with the reengineered SIPP. As long as it does, the consistent bias that, in theory, can be obtained with overlapping panels remains desirable. However, if the operational issues with overlapping panels were to persist, then, like the early SIPP, the benefits of overlapping panels would not be realized. Moreover, the atypically high poverty rates recorded in Wave 1 of the 2001 and 2004 panels, discussed in Chapter 2, present an additional complication. If this problem were to recur in the reengineered SIPP, it would diminish the value of combining estimates across panels, which would simply ensure that questionable estimates from Wave 1 were included in all pooled estimates. In light of these considerations, the panel does not recommend that overlapping panels be included in the initial design of the reengineered SIPP. At the same time, the panel underscores the importance of understanding panel bias and how it grows over time. Overlapping panels remain the surest way to document the extent of panel bias across the full range of variables collected in the survey. Unless the Census Bureau can find an alternative way to achieve this same result, it is important to conduct at least one pair of overlapping panels relatively early in the history of the reengineered SIPP.

CONTENT EVALUATION

A key element of reengineering SIPP and keeping it relevant to user needs concerns the survey content. In this regard, the panel commends the Census Bureau's efforts to reach out to the user community by asking users to comment on "content matrices" to help identify which portions of the survey are critical to users and to provide input on aspects of SIPP that

[1]Assuming funding for a fixed number of interviews, the number of panels that are in the field at the same time determines the average panel size. If two panels are in the field at the same time, then each panel can be only half the size that would be possible if a single panel were in the field at any one time.

could be improved. The degree of recent interaction between the Census Bureau and the SIPP user community is exemplary.

The charge to the panel did not include making specific recommendations on the content of SIPP. However, the panel thinks it important to comment on the need for a recurring, systematic review and evaluation of survey content to ensure that SIPP continues to serve its primary mission of providing data on the short-run dynamics of economic well-being. We also provide a discussion of the immigration data collected in SIPP as an example of the kind of assessment that the panel recommends.

An External SIPP Advisory Group

SIPP faces pressures to be everything to everyone. There is a perception among some inside and outside the Census Bureau that SIPP has been a dumping ground for new survey questions that various constituencies wish to see included in a Census Bureau survey. If true, the introduction of questions that are not essential to SIPP's core purpose is problematic, as the length and consequent respondent burden of the SIPP may have adverse implications for survey quality.

Any organization administering a major national survey must regularly assess survey content, focusing on three key questions: (1) Does there remain a compelling purpose for each question in the survey? (2) Is the question successfully gathering the information it is supposed to acquire at a reasonable cost? (3) Do the editing and imputation procedures for missing or erroneous responses reflect the insights of content experts?

To assist the Census Bureau in executing these three essential tasks for SIPP, the panel encourages the bureau to seek the expertise of content and survey specialists from government agencies, academic users, and policy analysis organizations by establishing a new advisory group. This group could be an expansion of the recently reconstituted SIPP Working Group sponsored by the American Statistical Association Survey Research Methods (ASA/SRM) Section, although we envision the charge to be broader than that working group's traditional mandate. Alternatively, it could be a new, separate, free-standing entity. It may be that federal government regulations will require that there be two groups—one group composed of federal agency staff, which could be the same as the interagency technical working group recommended in Chapter 3, and the other group composed of outside experts. Here we briefly discuss five tasks for the new advisory group(s).

First, the panel recommends that SIPP management staff and the advisory group make a periodic top-to-bottom review of SIPP survey content, ensuring that questions are consistent with SIPP's core mission and that each question is worth the cost—both the dollar cost and the opportunity cost in terms of questions forgone. Response burdens are high and survey space is precious. It is imperative that scarce questionnaire space be used effectively. The advisory group might also be helpful to the Census Bureau's efforts to resist adding content to SIPP that is not consistent with the survey's core mission.

Second, the SIPP management staff should seek the SIPP advisory group's thoughts on questionnaire changes that are likely to be made necessary by policy developments. When Aid to Families with Dependent Children was abolished, for example, or Medicare Part D was enacted, the SIPP questionnaire had to evolve. We expect that future changes in health insurance coverage and in the treatment of immigration will have implications for SIPP survey content. Members of the SIPP advisory group should have valuable content expertise to offer the Census Bureau staff responsible for meeting these and other evolving demands.

Third, with information provided by SIPP program staff, the SIPP advisory group should review the evidence on question nonresponse rates: At some threshold, whether 30 percent, 45 percent, 60 percent, or some other threshold, the information elicited by questions becomes worthless. If specific questions are generating low response rates, then additional work needs to be done to elicit higher response rates, or the questions should be dropped and replaced with values from administrative records or imputations when feasible. Work assessing response rates (and their threat to question integrity) should be augmented with a more targeted cognitive research program. It is likely that individuals have difficulty answering some types of questions. It is clear from evidence based on the Health and Retirement Study, for example, that people do not understand questions about whether their pension is a defined benefit or defined contribution type. As another example, many if not most recipients would be unlikely to accurately answer a question about whether they received an earned income tax credit, since more than 60 percent use a paid tax preparer to receive the credit. It would be valuable for the Census Bureau, with input from the SIPP advisory group, to maintain an ongoing, targeted cognitive research program on whether specific questions are eliciting accurate, useful responses.

Fourth, as discussed in Chapter 3, the advisory group can help evaluate the quality of survey responses relative to administrative data benchmarks and help assess the quality of SIPP imputations.

Fifth, the advisory group will be a valuable resource to provide advice on the core SIPP design issues. These include decisions about the sampling frame, the usefulness of dependent interviewing (this refers to the practice of reminding respondents of prior answers), the extent of and effects of seam bias, optimal numbers of interviews, optimal time between interviews, optimal length of interviews, effects of attrition, ways to reduce attrition and nonresponse at Wave 1, optimal recontact efforts, extent of item nonresponse, adequacy of weighting adjustments, imputation for unit and item nonresponse, efforts to facilitate timely release of the data, and data distribution mechanisms.

To summarize, the SIPP advisory group—the ASA/SRM working group with an augmented charge or a newly constituted group (or groups)—should scrutinize and conduct, on an annual or semiannual basis, evaluations of SIPP survey content. These evaluations should focus on improving survey questions, cognitive understanding of questions, response rates, benchmarking survey responses against external, reliable sources, and imputation and editing procedures. The group will provide a sounding board for the Census Bureau's plans to redevelop SIPP survey content. And the group should be useful as policy developments occur (such as changes in the nation's health insurance system) that require SIPP content to be altered.

Because within-panel changes to survey content are disruptive to users and data collectors, changes to SIPP content should occur at the beginning of a new panel, whenever possible. Well before the decision dates for an upcoming SIPP panel, funds should be provided for the Census Bureau to conduct comprehensive evaluations as outlined above and for the SIPP advisory group and the Census Bureau to convene a large-scale conference to review the latest evaluation research and to make suggestions for improvement.

Immigration Questions in SIPP

An example of the type of content evaluation that the SIPP advisory group might constructively undertake arises with immigration data, which have become increasingly important for determining eligibility for federal and state public assistance programs. SIPP collects detailed, time-varying data about key program eligibility criteria, including income, assets, employment, marital status, and custody of minor children. But, information about a person's immigration status and history—including visa status (i.e., the terms under which an immigrant was admitted to the United States), citizenship status, and duration of legal residence in the United States—is also needed for program eligibility determination. About 13 percent of the population are immigrants (from the 2007 American Community Survey, available at http://factfinder.census.gov), and roughly 20 percent of children have at least one immigrant parent (Urban Institute, 2006). Thus, the accurate determination of eligibility for immigrants is consequential, particularly for subgroups that contain large portions of immigrants, such as Hispanics and Asians.

During the 1980s and 1990s, the federal government and some states increased restrictions on immigrants' access to public resources. Unauthorized immigrants have long been excluded from eligibility from almost all public assistance programs except under emergency circumstances and when public health comes into play. In the early 1980s, Congress limited new immigrants' eligibility for public assistance during the 3 years following their official settlement in the United States, although refugees and asylees were excluded from these provisions, which required the use of visa status and duration of residence as eligibility criteria. Citizenship became important when Congress enacted the 1996 Personal Responsibility and Work Opportunity Act, when non-U.S. citizens were barred from many federal and some state programs. In general, eligibility for welfare was (and still is) linked to visa status (refugees versus legal permanent residents versus others), work history, and naturalization, although eligibility criteria for state-funded and some jointly funded federal and state programs differ substantially across states (Zedlewski and Giannarelli, 2001; Zimmerman and Tumlin, 1998).

SIPP currently collects information on immigration status that permits the rough approximation of eligibility on the basis of immigration-related criteria. Since the first SIPP panel in 1984, data on place of birth, period of entry (year first moved to the United States), a partial migration history (place of previous residence and timing of moves to current and previous residence), and citizenship status have been collected from all adults ages 15 and older in a topical module, usually as part of Wave 2. Starting with the 1996 panel, limited information on visa status was added, including information about whether the respondent was originally admitted as a legal permanent resident, and whether and when the person subsequently converted to this status. In addition, the core questions in each wave include for both adults and children under age 15 whether the person was born outside the United States, citizenship status, and how the person became a citizen (e.g., through birth or naturalization).

The inclusion of a partial migration history in the Wave 2 topical module not only provides valuable information about internal migration, but also can be used to supplement information on the timing of international migration. For example, although the data on period of entry include many missing values (about 25 percent of the foreign-born in the 2004 SIPP panel were coded as missing this item), it is possible to use other information from the migration history to identify periods of time when the respondent lived in the United States and thus identify those who were living in the country long enough to be eligible for certain programs. Of the foreign-born with missing data on period of entry, 65 percent provided information on the timing of moves into current or previous U.S. residences, and, of these, 63 percent indicated they had been living in the United States at least since 1990; 85 percent had been living in the United States at least since 2000.

The Census Bureau should be commended for attempting to collect information on migration history and immigration status. Immigration status can limit (or enhance) people's opportunities and is an important criteria for program eligibility. No other nationally representative survey contains information on the status of immigrants upon entry to the United States (the New Immigrant Survey contains detailed information on the immigration and admission status of legal immigrants, but not unauthorized or nonimmigrants—see http://nis.princeton.edu). In addition, SIPP is the only nationally representative population sample that follows a large sample of immigrants over time. In these respects, SIPP is a unique, valuable data source for immigration scholars. However, additional information would further enhance the usefulness of the data for policy-relevant analysis of immigrant populations. Three specific suggestions are listed below.

Ask migration history questions for new adult household members Currently, the detailed immigration information is asked only in Wave 2. To obtain a complete picture of the migration history of household members, it would be useful to administer the migration history questionnaire to adults who join a sample household after Wave 2.

Collect information on parents' place of birth A major question about immigrant populations concerns the degree to which they change and adapt with increasing time in the country. Duration in the country can be measured as time since arrival within the lifetime of immigrants themselves or as the number of generations a person's family has been in the country (i.e., first-generation immigrants, second-generation U.S.-born children of immigrants, and third-or-higher generation U.S.-born children of U.S.-born parents). Although SIPP includes information about the timing of immigration for individuals, it would be useful to also collect data on mother's and father's place of birth, which would permit the identification of the first, second, and third-or-higher generations. Currently, the monthly CPS is the only nationally representative sample that includes information on parents' place of birth. The addition of these items to SIPP would make it possible to compare income dynamics and other characteristics of immigrant generations.

Investigate alternative techniques for collecting sensitive information on immigration status By collecting data on immigration status, SIPP goes well beyond most other surveys. Nevertheless, the quality of the data on immigration status is questionable. Many respondents fail to answer these questions, and, of those who do, many appear to provide inaccurate information. Among the foreign-born in the 2004 panel migration history topical module, 28 percent did not answer the question about immigration status (compared with 21 percent for the question on country of birth). In addition, the accuracy of reporting is doubtful. For example, among Mexican-born adults in the 2004 SIPP panel who reported on immigration status, 33 percent (weighted) said they were not admitted as a legal permanent resident, had not naturalized, and had not converted to this status, thus suggesting that no more than 33 percent were unauthorized. But other estimates based on demographic methods suggest that nearly half (47 percent) of the Mexican foreign-born were unauthorized migrants in 2004 (Passel, 2006). The imputation procedures used in SIPP to fill in missing values do not improve the situation. When imputed responses are included in the sample, the upper-bound estimate of unauthorized migrants drops to 28 from 33 percent.

It is understandable that many unauthorized migrants would misreport their citizenship or immigration status to employees of the U.S. federal government. One possible way to improve reporting is to use a self-administered questionnaire for these items. Another possibility is to use the randomized response method, first introduced by Warner (1965).[2]

Still another way to improve the accuracy of data on immigration status is to attempt to match respondents with the immigration admission and naturalization administrative records of the Office of Immigration Statistics (OIS) in the U.S. Department of Homeland Security. Matching these data would be challenging because the electronic OIS records currently do not contain a field for Social Security number (personal communication with OIS). Thus, matches would have to be made on the basis of such identifiers as name, sex, date of birth, year of admission, and country of birth, although the Census Bureau has made striking advances in its ability to link data based on these or similar characteristics.

If SIPP foreign-born respondents were successfully matched to OIS admission and naturalization records, the information in the administrative records could be used to improve the quality of SIPP data on citizenship and immigration status. For example, matched data could be used to evaluate the accuracy of responses generated by alternative survey methodologies (e.g., in-person interviews versus self-administered questionnaires, or the random response method versus standard questions). In addition, matched data could be used to improve imputations of missing data on immigration and citizenship status as well as items related to immigration status—for example, unauthorized immigrants are ineligible for many public assistance programs, so they should not be imputed as recipients.

[2]Respondents are presented with two alternative questions—one about their immigration status and another on an innocuous topic (e.g., favorite color). Respondents then roll a die in private (or engage with some other random device) to determine which question to answer (e.g., those rolling "1" or "2" answer the question about favorite color, and those rolling other numbers answer the question about immigration status). Because no one but the respondent knows which question was answered, privacy is maintained, and respondents may be more likely to give truthful answers. Response error is better managed because it is more likely to be randomly distributed. Statistical methods have been developed for analyzing this type of data. See also U.S. General Accounting Office (1999), which proposes a three-card method for collecting sensitive information such as immigration status.

SIPP PROCESSING, ACCESS, MANAGEMENT, AND BUDGET

Timeliness

Given the absence of an external agency SIPP sponsor (discussed below), it is critical that SIPP meet the needs of its large, diverse user community in order to have a strong base of support. The panel thinks SIPP data would be used even more extensively if the Census Bureau could significantly shorten the amount of time needed to release the data, consistent with maintaining a high-quality product. One model for efficiency of data collection and release in the Census Bureau itself is the CPS. For example, data from the CPS Annual Social and Economic Supplement (ASEC) (which are typically collected in February, March, and April) are made publicly available by August of the same year.

There are several reasons why this time frame could not realistically be applied to SIPP, a prominent one being that processing longitudinal SIPP data is in many ways considerably more complicated than processing the cross-sectional information collected in the CPS ASEC supplement. The SIPP instrument is also longer and collects a broader range of information. Nonetheless, as noted in Chapter 2, the release of SIPP data is often not timely, lagging 2 or more years behind data collection.

One survey that is more comparable to SIPP than the CPS is the PSID. Like SIPP, the PSID is a longitudinal household survey that asks a broad array of questions on demographics and employment. The PSID has the advantage of going into the field every 2 years, rather than every 4 months, as SIPP does. The PSID generally releases data in a timelier manner than SIPP—typically 12 months after the data are collected. The Medical Expenditure Panel Survey also releases each year "point-in-time" public-use files within 12 months of data collection; these files are based on a single round of interviewing (from two overlapping panels) and in that respect are similar to SIPP wave files. A reasonable goal for the reengineered SIPP to adopt could be to release wave files within 12 months of data collection, and, indeed, the SIPP 2001 panel data were released on roughly this schedule.

The usefulness of SIPP data to users would be increased by consistently having a relatively short lag time between data collection and release of 1 year or less. The Census Bureau is capable of timely dissemination of data, as evidenced by the efficiency of the processing of the CPS ASEC supplement and occasional past SIPP panels. The bureau needs to ensure that the same type of management attention and coordination is applied to ensure timely delivery of future SIPP panels, particularly in years when the survey instrument or processing procedures are being updated, which occurs periodically.

The panel anticipates that the move to the BLAISE-based instrument and SAS-based processing system will improve the speed at which the reengineered SIPP is processed. Regardless, the Census Bureau should identify the key bottlenecks that are hindering timely release of the data and take the steps necessary to reduce them, while not forgoing thorough quality checks that might help prevent the need to rerelease a SIPP file with corrections. The goal should be to meet the best practices of other national surveys in the release of data. The panel thinks that 1 year between the end of a survey and data release should be an achievable target.

Enhancing Access to SIPP

One common complaint from current and prospective SIPP data users is the difficulty associated with working with SIPP files. Longitudinal files are inevitably more complex than cross-sectional files, particularly for researchers interested in linking individual and household information over time. Moreover, since each wave of a SIPP panel consists of four staggered rotation groups, new users often grapple with creating calendar-year files (if that is their goal). Most importantly, the quality and quantity of documentation of SIPP files was poor in the past.

SIPP documentation is improving. An early edition of a *SIPP Users' Guide* was released in 1987 and updated in 1991. A comprehensive third edition was released in 2001 (available at http://www.census.gov/sipp/usrguide.html), which is currently being updated to include information about the 2001, 2004, and 2008 panels. The SIPP website also provides a link to a tutorial (see http://www.census.gov/sipp/). Moreover, in recent years, it has become easier to access and download SIPP data over the Internet. The main mechanisms for downloading SIPP data from the Census Bureau are via (1) a file transfer protocol (FTP) with a link at the SIPP home page, which is for users who wish to download entire longitudinal, core, or topical module files and (2) the DataFerrett application tool, with which researchers can download a subset of variables or observations from particular SIPP files.

Despite documentation improvements and the various data extraction tools available, there is still room for improvement. For example, a rather minor change would be to integrate the documentation that is available at the SIPP homepage with the DataFerrett data extraction tool. The latter could at least have various links to the former. More importantly, the process of updating the *SIPP Users' Guide* should be completed as soon as possible. Chapters of the guide that have not yet been revised refer only to data up to the 1996 panel. Another feature that would assist some users would be to provide code on how to construct calendar-year files, which would assist them in dealing with the complexities introduced by having different rotation groups for a given wave. This issue would become irrelevant, of course, if the SIPP moves to the EHC instrument that collects data annually, as the rotation groups would be eliminated. Finally, the Census Bureau could enhance DataFerrett, making it even easier to use (see Box 4-1).

BOX 4-1
Improving Access to SIPP Data via DataFerrett

DataFerrett (available at http://dataferrett.census.gov/) is the central access point for many users to data from the Survey of Income and Program Participation. It is an online data access tool that permits users to create a customized data extract of selected variables and observations from any one of a large number of Census Bureau data sets, including SIPP. The user interface of DataFerrett is "point-and-click" and does not require specialized programming knowledge. Users are guided through several steps in which they select a data set (e.g., the 2001 SIPP longitudinal file), select a set of variables from the data set, and select a subsample (e.g., men ages 20-29). Users then may either download the data extract (so that they can analyze it with their own statistical software) or continue to work online to create a table of descriptive results (e.g., frequency distributions, cross-tabulations). Several points of concern about DataFerrett warrant further scrutiny by the Census Bureau to improve access to SIPP and other data sets:

- Tutorial—In general, the directions in the tutorial for using DataFerrett are unclear.
- SIPP-specific information—DataFerrett is not tailored for any specific data set; the user interface and information provided are structured in the same way for the Current Population Survey, the American Community Survey, and SIPP. Yet there are unique features of SIPP that may require special treatment. For example, SIPP is longitudinal, and data for each panel are contained in several files, which may not be readily apparent to a new user. Another unique feature of SIPP is its topical modules. Although DataFerrett will display information about each data set, the specific information provided about the contents of the topical modules is not useful.
- Variable selection—Finding and selecting variables in DataFerrett can be tedious and frustrating. For example, once users have selected a list of variables, they always have to click the browse/selection variables and values button, then click the selection box, then click ok. An easier approach should be possible. The search tool for variable selection could be improved by providing an "advanced search" option in which users can enter four or five search items and combinations of those items (using either AND or OR), and by providing a list of commonly used search terms or list of variables or topic areas.

SIPP Management and Budget

As we recounted in Chapter 2, SIPP has a unique position among the Census Bureau's data creation activities for the household sector. Unlike other surveys of people and households that the Census Bureau conducts, SIPP does not have a government client outside the Census Bureau or a federally mandated set of reports that are based on the survey. The earlier

It would be helpful if DataFerrett provided more guidance to users about which variables to include in their data extracts. First-time users (and even experienced analysts) may be confused about or unaware of important variables to include, such as sampling weights and key identifiers (e.g., sampling unit, address, family, person, entry identification, and wave). DataFerrett could provide a description of these key variables and alert users if they fail to download them. Other data access programs—such as the one used with the Integrated Public Use Microdata Series (IPUMS; see http://www.ipums.umn.edu)—go so far as to automatically include these key variables on all extracts.

- Merging data across waves—One of the barriers for new users in working with SIPP is its complex, longitudinal design. DataFerrett could be designed to provide an easy-to-use, transparent way of merging data for individuals across waves. One especially valuable feature—the ability to select and download in a single extract variables from multiple topical module and core data files and waves across a panel—exists but is very hard to find in the current interface. Also, the task of selecting variables from multiple data files (e.g., from a topical module and the core) can be tedious. A better design might be to list all of the variables in the core and topical modules together in one place (not broken down by data file or wave). As the user selects variables, information on the available waves for the selected variable would pop up, and the user would then select the waves he or she wants. This design would make it easier to quickly identify and download all variables that repeat across waves of a panel and would not require users to know in advance which items are in which topical modules.
- Table and recode functions—The tabulation and recode functions are difficult to determine how to use, and some users may not find them helpful. It is difficult to code a variable as a dummy or to assign the same value to more than one variable. In addition, DataFerrett does not permit users to export tables as Microsoft Excel files. It would be helpful to include a prominent button that users can select if they want to export a table. A dialog box could then appear with various format options, including an Excel worksheet.

SOURCE: Analysis by students of panel member Jennifer Van Hook.

Committee on National Statistics SIPP panel recommended that this situation be addressed, most naturally by making a required report to Congress on poverty (or poverty transitions) based on SIPP (National Research Council, 1993:85). This recommendation was not adopted. Not having an external client, such as the Bureau of Labor Statistics (which has a collaborative and financial stake in the monthly CPS), or a set of regular reporting requirements, as with the decennial census and the American Community

Survey, has contributed to setbacks in the development of SIPP (see also National Research Council, 2001:150-154, on this point). In addition, as described in Chapter 2 and in the prior SIPP report (National Research Council, 1993:20), the value of the survey has been materially diminished over its history by sample cutbacks necessitated by cutbacks in funding.

Historically, SIPP has also lacked a project director with full management and budget authority for all aspects of the survey. A recommendation in the earlier SIPP report reads as follows (National Research Council, 1993:235-236):

> To be as effective as possible in carrying out its responsibilities to produce timely, comprehensive, relevant, high-quality, and analytically appropriate statistics on income and program participation, the Census Bureau should establish a senior-level position of project director for the Bureau's income surveys, SIPP and the March CPS income supplement. This position should include full management and budgetary authority for the income statistics program and sufficient resources to obtain the level of analysis staff that is needed to provide substantive guidance to the program, prepare reports, conduct analyses, and evaluate analytical concepts and methods. The person who fills this position should have recognized substantive expertise in topics related to income, poverty, and assistance programs, combined with strong survey management skills.

This recommendation was never acted upon, yet we continue to think that SIPP would benefit from a project director with a distinct budget. The budget must always include adequate research and development funding, since SIPP is a major ongoing survey that requires regular evaluation and improvement.

CONCLUSIONS AND RECOMMENDATIONS

Event History Calendar Approach

Conclusion 4-1: The Survey of Income and Program Participation (SIPP) is the only national survey that provides information on the short-term dynamics of employment, income, program participation, and other family characteristics, and its monthly time frame is essential for many applications. The Census Bureau's plans to move SIPP to an annual survey, filling in intrayear dynamics using event history calendars, potentially affects—perhaps positively, perhaps negatively—SIPP's single most important feature.

Conclusion 4-2: The panel is not aware of conclusive evidence that a 12-month event history calendar (EHC) framework is capable (or not) of generating accurate monthly information on income, program participation, and other topics that are covered in the Survey of Income and Program Participation (SIPP). The lack of evidence about the ability of an EHC to collect monthly data places considerable pressure on the Census Bureau, not only to design an effective pretesting program for the EHC methodology, but also to make its survey reengineering plans for SIPP sufficiently flexible so that it can modify its plans if the pretesting reveals unanticipated, negative evidence on the likely success of the proposed methodology in providing high-quality monthly information.

Conclusion 4-3: Understanding transitions at the seam between interviews in a reengineered Survey of Income and Program Participation (SIPP) using the event history calendar approach will require data from at least two annual interviews. Moreover, not enough is yet known about the factors driving seam bias in the traditional SIPP.

Conclusion 4-4: A parallel traditional Survey of Income and Program Participation (SIPP) panel that provides 2 or more years of data is a necessary component of a thorough evaluation of the reengineered SIPP using the event history approach. The recently completed paper test is of limited value for this purpose. The Census Bureau's planned electronic prototype test is promising, but, as a single test, is unlikely to provide conclusive findings.

Recommendation 4-1: The Census Bureau should engage in a major program of experimentation and evaluation of the event history approach for developing suitable data on the short-run dynamics of household composition, income, employment, and program participation from a reengineered Survey of Income and Program Participation (SIPP). The details of the Census Bureau's plans should be disseminated to SIPP stakeholders for comment and suggestions for improvement. If the experimental results indicate that the quality of data on income and program dynamics is significantly worse under the event history calendar approach than in the traditional SIPP, the Census Bureau should return to a more frequent interview schedule, say, every 6 months, devise other methods to improve data on short-run dynamics, or revert to the traditional SIPP with 4-month interviews using standard questionnaires.

Recommendation 4-2: To ensure not only adequate evaluation of a reengineered Survey of Income and Program Participation (SIPP), but also a bridge between data collected under the new and old methods, the Census Bureau should conduct traditional and reengineered SIPP panels to provide at least 2 years of comparable data. If the new design works, then the parallel traditional panel provides a bridge. If the new design does not work, then the parallel panel provides a backup for the continued collection of SIPP data while the new design is modified as appropriate.

Recommendation 4-3: Because the reengineered Survey of Income and Program Participation (SIPP) should be compared with the first year of a traditional SIPP panel in order to minimize attrition bias, the Census Bureau should begin a new traditional SIPP panel in February 2012. If the costs of fielding two concurrent national longitudinal surveys appear prohibitive, the 2012 traditional SIPP panel could be smaller than previous SIPP panels without substantially diminishing its scientific value.

Length and Frequency of Interviews and Panels

Conclusion 4-5: Design features for a reengineered Survey of Income and Program Participation (SIPP) that are important to evaluate in terms of their effects on respondent burden, survey costs, data quality, and operational complexity include the length and frequency of interviews, the length of panels, and whether successive panels overlap. With regard to interviews, there is no evidence that a 12-month event history calendar strikes the optimal balance between respondent burden, costs, and data quality in comparison to the traditional SIPP design of 4-month interviews. With regard to panels, there is evidence that nonoverlapping panels have adverse effects on cross-sectional estimates of trends over time, yet they are advantageous in terms of larger sample sizes per panel and operational feasibility.

Recommendation 4-4: The Census Bureau should study the trade-offs in survey quality and respondent burden in comparison to survey costs between longer but less frequent event history-based interviews in a reengineered Survey of Income and Program Participation (SIPP) and more frequent interviews in the traditional SIPP. The Census Bureau's research and evaluation program for SIPP should also improve understanding of panel bias and how it grows over time. Because overlapping panels remain the best way to document the extent of panel bias across the full range of variables collected in SIPP, they should be on the research agenda for possible implementation at a future time. Due to technical demands and capacity issues that arise in launching the reengineered SIPP, the initial design plans should not include overlapping panels.

Content

Conclusion 4-6: The Census Bureau has done an exemplary job in reaching out to the Survey of Income and Program Participation user community with "content matrices" and other efforts to identify critical portions of the core questionnaire and topical modules for data users.

Recommendation 4-5: The Census Bureau should expand the scope of the reconstituted Survey of Income and Program Participation (SIPP) Working Group or establish a new SIPP advisory group with members from academic institutions and policy research organizations that would meet periodically to assist the Census Bureau in its efforts to continually improve the quality and relevance of the SIPP survey content. This group, which could include government members from the recommended interagency working group on uses of administrative records in SIPP (see Recommendation 3-5), would review the Census Bureau's use of cognitive and other methods to evaluate and improve survey question wording and improve response rates (or, when that is not possible, either dropping the question or seeking an alternate data source); assist in benchmarking survey responses against external, reliable sources; and advise the bureau on ways to improve imputation and editing procedures. The group would provide a sounding board for the Census Bureau's plans to develop appropriate survey content in a reengineered SIPP and advise the bureau on appropriate modifications to survey content as policy developments occur, such as health care and immigration reform

Timeliness

Conclusion 4-7: The release of Survey of Income and Program Participation (SIPP) data is often not timely. Data from the 2004 SIPP panel were generally released more than 2 years after being collected. Other panel surveys have more timely data release, often within a year of data collection, which enhances their usefulness to external users.

Recommendation 4-6: The Census Bureau should release Survey of Income and Program Participation data within 1 year of data collection.

Management and Budget

Conclusion 4-8: Unlike other surveys of people and households that the Census Bureau conducts, the Survey of Income and Program Participation (SIPP) does not have a government client outside the Census Bureau or a federally mandated set of reports that are based on the survey. Not having an external client, such as the Bureau of Labor Statistics (which has a collaborative and financial stake in the monthly Current Population Survey), or a set of regular reporting requirements, as with the decennial census and the American Community Survey, has contributed to setbacks in the development of SIPP. The value of the survey has also been diminished over its history by sample cutbacks necessitated by cutbacks in funding.

We agree with an earlier Committee on National Statistics panel (National Research Council, 1993) that SIPP would benefit from a project director with full management and budget authority for design, evaluation, and operations. The budget should always include adequate research and development funding, since SIPP is a major ongoing survey that requires regular evaluation and improvement.

Appendix A

SIPP Data Quality

John L. Czajka

This appendix provides brief summaries of what is known about the quality of data from the Survey of Income and Program Participation (SIPP) in areas that are central to the survey's principal purposes or major uses. Topics include the following:

- Income
- Program Participation
- Income Receipt from Multiple Sources
- Wealth
- Health Insurance Coverage Transitions
- Attrition
- Representation of the Population Over Time
- Seam Bias
- Imputation
- Wave 1 Bias

These topics are discussed in the order that they are listed.

INCOME

In comparison with the Annual Social and Economic Supplement (ASEC) to the Current Population Survey (CPS), the official source of income and poverty statistics for the United States, SIPP captures nearly as much transfer income and substantially more self-employment income but less wage and salary income and substantially less property income. These last two sources dominate earned and unearned income, respectively; as a result, SIPP underestimates total CPS income by 11 percent according to a recent comparison based on calendar year 2002 (Czajka and Denmead, 2008). This underestimation reflects a deterioration in the relative quality of SIPP income data since the survey's inception.

Early SIPP

Comparisons of income estimates from the first SIPP panel with the CPS and independent benchmarks were quite favorable to SIPP. In its estimate of aggregate income for calendar year 1984, SIPP captured 99.9 percent as much regular money income—that is, excluding lump sums—as the CPS (Vaughan, 1993). SIPP captured nearly 12 percent more transfer income—a major focus of the survey—and 3 percent more property income. Relative to independent estimates from program administrative data, SIPP captured 101 percent of aggregate Social Security income, 98 percent of Supplemental Security Income (SSI), 82 percent of Aid to Families with Dependent Children benefits, 96 percent of general assistance benefits, 77 percent of veterans' compensation or pension income, and 87 percent of unemployment compensation. SIPP estimates of aggregate pension dollars by type were between 95 and 103 percent of independent estimates. However, SIPP's estimate of total earnings, the largest component of total income by far, was 1.8 percentage points below the CPS. Furthermore, SIPP's shortfall on earned income was the net result of differential performance for wage and salary employment and self-employment. SIPP's estimate of self-employment income exceeded the CPS estimate by 45 percent, but for wage and salary income SIPP captured 5.3 percent fewer total dollars than the CPS. Relative to an independent estimate from the national income and product accounts (NIPAs), the CPS captured 98 percent of total wage and salary income and SIPP captured 92.6 percent.

SIPP's success with self-employment income was the result of a nonconventional measurement approach that rejected the traditional definition of such income as revenue less expenses (or profit/loss). The SIPP approach grew out of efforts to translate the conventional approach to a subannual reference period, during which revenues and expenses might fluctuate widely—if they were known at all. SIPP staff sought a better measure of the income that business owners obtained from their businesses on a month-to-month basis. Rather than asking about profit and loss, SIPP asks respondents how much they withdrew from each business during each month of the reference period. One consequence of this approach is that self-employment income cannot be negative in SIPP.[1] In the CPS in the mid-1980s, roughly a fifth of the self-employed reported net losses from their businesses.

With respect to wage and salary income, SIPP's shortfall occurred despite the survey's finding 1.3 percent more workers than the CPS. The composition of the workers identified by SIPP may have contributed to the difference in aggregate dollars. Compared with the CPS, SIPP found 13 percent more workers who were employed less than full-time, full-year, but 7 percent fewer full-time, full-year workers. SIPP's success in finding part-time and part-year workers seemed to be a direct result of the survey's more frequent interviews and shorter reference periods relative to the annual interviews and annual reference period of the CPS. The smaller number of full-time, full-year workers in SIPP could also have reflected a more accurate reporting of hours and weeks worked. If that were the case, however, the lower aggregate income obtained in SIPP would have been due entirely to workers reporting lower income from their employment than workers responding to the CPS.

SIPP Income Over Time

Between 1984 and 1990, the SIPP estimate of total income slipped below 98 percent of the CPS aggregate according to analyses reported by Coder and Scoon-Rogers (1996) and Roemer (2000).[2] This reduction was distributed across a large number of income sources, with no single source or small number of sources being primarily responsible for the change.

---

[1]In the 2004 panel, SIPP started to ask separately for the amount of profit or loss over the 4-month reference period and to include this amount in monthly income totals. Net negative income from self-employment—not previously provided in the SIPP public-use files—will now be provided.

[2]To estimate aggregate annual income with SIPP, one must sum the monthly amounts reported by respondents who may not have been present—in the sample or even in the population—for the entire calendar year. There are different ways to do this, and they vary with respect to which months are counted for which persons and what weights are applied to them. Coder and Scoon-Rogers (1996) describe three methods and provide SIPP estimates for all three. None of the three methods is inherently more valid than the others; they just represent different ways of looking at the income data collected by SIPP, although two of the methods are more consistent with the way that SIPP collects income data. The third method, which is designed to resemble the CPS, requires an adjustment for missing months. The first method, which sums the monthly aggregates for all respondents present each month, makes the fullest use of the income data reported for a calendar year, but it yields slightly lower annual income estimates than the other two methods for 1990. Coder and Scoon-Rogers used the third method, and Vaughan (1993) and Roemer (2000) used the first method.

Between 1990 and 1996, a period that saw the introduction of computer-assisted interviewing to both surveys and a major redesign of SIPP, Roemer's (2000) detailed analysis shows that SIPP fell further behind the CPS. Relative to independent NIPA benchmarks, SIPP estimates of total income dropped only slightly, from 87.1 percent in 1990 to 85.7 percent in 1996 (see Table A-1). More substantial reductions were recorded for property income (9 percentage points) and transfers (6 percentage points). Estimates of pension income increased by a percentage point, as did wages and salaries, but income from self-employment fell from 85 percent of the benchmark in 1990 to 69 percent by 1996. Given that the SIPP concept of self-employment income differs from the conventional concept, the decline should probably be attributed to a growing gap between the two concepts rather than anything in the survey. Finally, it may reflect favorably on some aspects of the SIPP redesign that the estimate of SIPP total income relative to the benchmark rose by a percentage point between 1995 and 1996 after having declined from 87.1 percent in 1990 to 84.8 percent in 1994 and 1995.

Over the same period, however, CPS total income increased by 3 percentage points relative to the benchmark (see Table A-2). CPS estimates of wages and salaries increased from 95.9 to 101.9 percent of the NIPA estimate; property income rose from 62.8 to 70.0 percent; and transfer income increased from 87.6 to 88.3 percent. Pensions declined, however, from 88.9 to 76.6 percent, and self-employment income dropped from 68.5 to 52.6 percent. The biggest increase occurred between the 1992 and 1993 reference years, which coincided with the introduction of computer-assisted interviewing in the CPS. One element of the switch from a paper and pencil instrument was clearly related to the increased amount of income collected: the maximum amount of wage and salary income that could be reported was increased from $499,997 to $2,099,999. Roemer determined that this change alone added 2 percentage points to the CPS income total relative to the NIPA total.[3]

The combined impact of the SIPP and CPS changes over this period was to reduce the ratio of SIPP to CPS total income to 92.5 percent (see Table A-3). Wages and salaries in SIPP dropped from 94.0 to 89.3 percent of the CPS estimate, although self-employment income increased by 7 percentage points relative to the CPS. SIPP property income fell substantially, going from 104.0 to 80.9 percent of the CPS. Even transfer income dropped from 105.0 to 97.7 percent of the CPS estimate, but this could be attributed primarily to Social Security income, which fell from 105.6 to 98.4 percent of the CPS estimate between 1993 and 1994. The shift between those two years was owing to an increase in the amount reported in the CPS rather

[3]There does not appear to have been a similar issue with respect to the collection of wage and salary income in SIPP, given that annual earnings are constructed from monthly earnings.

TABLE A-1 Survey Income as a Percentage of Independent (NIPA) Benchmarks: SIPP, 1990 to 1996

	Survey Reference Year						
Income Source	1990	1991	1992	1993	1994	1995	1996
Total Income	**87.1**	**87.9**	**84.9**	**86.9**	**84.8**	**84.8**	**85.7**
Earnings	**89.6**	**90.9**	**86.9**	**87.4**	**86.4**	**86.7**	**88.4**
Wages and salaries	90.1	90.5	88.1	89.0	88.5	88.3	91.0
Self-employment	85.1	94.6	77.7	76.2	70.5	75.0	69.1
Property Income	**65.3**	**60.2**	**60.5**	**77.0**	**60.1**	**58.9**	**56.6**
Interest	56.7	56.6	56.5	62.1	51.3	51.3	50.2
Dividends	65.8	53.3	50.5	95.9	62.5	65.8	51.0
Rent and royalties	113.1	90.7	90.8	91.2	81.0	69.2	82.0
Transfers	**92.0**	**90.5**	**89.0**	**89.4**	**87.8**	**87.0**	**86.3**
Social Security and Railroad Retirement	97.1	95.0	93.6	92.7	90.8	90.9	87.9
Supplemental Security Income	83.1	88.6	84.9	82.9	86.0	86.2	101.4
Family assistance	75.6	76.4	69.9	89.1	87.3	85.8	76.3
Other cash welfare	81.9	100.9	81.3	96.6	79.2	95.9	114.0
Unemployment compensation	77.5	83.5	82.4	86.3	84.3	75.7	69.4
Workers' compensation	67.8	61.5	68.6	59.2	57.8	51.2	71.7
Veterans' payments	83.1	78.8	79.5	77.5	75.6	72.7	72.9
Pensions	**84.6**	**87.9**	**84.9**	**86.9**	**84.8**	**84.8**	**85.7**
Private pensions	91.8	85.7	86.7	96.9	103.8	99.5	98.1
Federal employee pensions	75.9	89.8	84.6	86.3	89.0	88.5	75.6
Military retirement	87.4	92.0	83.4	87.3	87.1	85.4	101.6
State and local employee pensions	76.8	84.2	80.1	76.6	77.0	74.3	67.8

NOTE: Survey estimates are based on the Census Bureau's internal data, without top-coding; however, there are limits on the amount of income that can be reported, which vary by source.
SOURCE: Roemer (2000:Table 3b); data from the 1990, 1991, 1993, and 1996 SIPP panels.

TABLE A-2 Survey Income as a Percentage of Independent (NIPA) Benchmarks: March CPS, 1990 to 1996

	Survey Reference Year						
Income Source	1990	1991	1992	1993	1994	1995	1996
Total Income	**89.3**	**89.4**	**88.0**	**91.7**	**92.9**	**92.2**	**92.6**
Earnings	**93.0**	**93.0**	**91.3**	**94.8**	**96.4**	**95.1**	**96.1**
Wages and salaries	95.9	96.4	95.6	99.7	101.9	101.4	101.9
Self-employment	68.5	65.3	58.6	58.9	54.8	48.5	52.6
Property Income	**62.8**	**63.6**	**63.2**	**69.8**	**65.7**	**72.9**	**70.0**
Interest	67.1	68.3	67.6	79.7	72.3	83.9	83.8
Dividends	40.9	45.7	49.2	54.3	54.6	62.6	59.4
Rent and royalties	85.0	74.1	69.8	65.2	64.8	58.7	58.6
Transfers	**87.6**	**86.8**	**83.6**	**85.6**	**89.5**	**89.2**	**88.3**
Social Security and Railroad Retirement	90.6	88.6	87.1	87.8	92.3	92.0	91.7
Supplemental Security Income	78.9	84.6	75.5	84.2	78.0	77.1	84.2
Family assistance	74.4	74.4	72.2	76.4	73.1	70.5	67.7
Other cash welfare	85.6	77.5	81.6	101.3	105.2	95.8	80.5
Unemployment compensation	79.9	82.5	72.8	77.6	90.0	91.3	81.6
Workers' compensation	89.5	89.1	82.5	77.0	77.7	69.3	62.7
Veterans' payments	73.9	82.9	77.7	85.5	84.7	94.9	89.6
Pensions	**88.9**	**85.5**	**83.1**	**83.6**	**83.1**	**78.2**	**76.6**
Private pensions	98.3	96.3	96.4	98.8	102.7	93.9	93.1
Federal employee pensions	82.7	82.6	84.5	82.7	80.9	77.9	80.8
Military retirement	85.6	84.6	74.3	71.7	76.4	70.6	58.2
State and local employee pensions	78.7	68.5	64.2	66.7	59.6	59.0	57.3

SOURCE: Roemer (2000:Table 2b); data from the 1991 through 1997 ASEC supplements to the CPS.

TABLE A-3 SIPP Aggregate Income as a Percentage of March CPS Aggregate Income, 1990 to 1996

	Survey Reference Year						
Income Source	1990	1991	1992	1993	1994	1995	1996
Total Income	**97.5**	**98.3**	**96.5**	**94.8**	**91.3**	**92.0**	**92.5**
Earnings	**96.3**	**97.7**	**95.2**	**92.2**	**89.6**	**91.2**	**92.0**
Wages and salaries	94.0	93.9	92.2	89.3	86.8	87.1	89.3
Self-employment	124.2	144.9	132.6	129.4	128.6	154.6	131.4
Property Income	**104.0**	**94.7**	**95.7**	**110.3**	**91.5**	**80.8**	**80.9**
Interest	84.5	82.9	83.6	77.9	71.0	61.1	59.9
Dividends	160.9	116.6	102.6	176.6	114.5	105.1	85.9
Rent and royalties	133.1	122.4	130.1	139.9	125.0	117.9	139.9
Transfers	**105.0**	**104.3**	**106.5**	**104.4**	**98.1**	**97.5**	**97.7**
Social Security and Railroad Retirement	107.2	107.2	107.5	105.6	98.4	98.8	95.9
Supplemental Security Income	105.3	104.7	112.5	98.5	110.3	111.8	120.4
Family assistance	101.6	102.7	96.8	116.6	119.4	121.7	112.7
Other cash welfare	95.7	130.2	99.6	95.4	75.3	100.1	141.6
Unemployment compensation	97.0	101.2	113.2	111.2	93.7	82.9	85.0
Workers' compensation	75.8	69.0	83.2	76.9	74.4	73.9	114.4
Veterans' payments	112.4	95.1	102.3	90.6	89.3	76.6	81.4
Pensions	**95.2**	**102.8**	**102.2**	**103.9**	**102.0**	**108.4**	**111.9**
Private pensions	93.4	89.0	89.9	98.1	101.1	106.0	105.4
Federal employee pensions	91.8	108.7	100.1	104.4	110.0	113.6	93.6
Military retirement	102.1	108.7	112.2	121.8	114.0	121.0	174.6
State and local employee pensions	97.6	122.9	124.8	114.8	129.2	125.9	118.3

SOURCE: Tables A-1 and A-2.

than a decline in what was reported in the SIPP.[4] However, later analyses of SIPP data matched to Social Security administrative records uncovered a tendency for respondents to report their Social Security payments net of their Medicare Part B premiums, which are deducted from their monthly benefit checks or automated payments (Huynh, Rupp, and Sears, 2001). In an apparent concession to respondents, the SIPP instrument was changed after the first wave of the 1993 panel to explicitly request that Social Security benefits be reported net of the Medicare premiums. The SIPP instrument was revised again for the 2004 panel to collect the amount of the Medicare premium as a separate quantity, which the Census Bureau could then add to the reported net payment to obtain the gross amount. Finally, SIPP pension income increased from 95.2 to 111.9 percent of the CPS estimate due to the decline in pension dollars collected in the CPS.

Quality of Wage and Salary Data

To gain a better understanding of the biggest source of the discrepancy between SIPP and CPS total income, Roemer (2002) compared both SIPP and CPS annual wages and salaries to the wages and salaries reported in the Social Security Administration's Detailed Earnings Records (DER) for 1990, 1993, and 1996. Unlike other Social Security wage records, the DER is not capped at the income subject to the Social Security tax, and unlike tax records it includes deferred compensation. Roemer's comparisons used survey records that had been matched to the DER based on the Social Security numbers reported by SIPP and CPS respondents, allowing an assessment of discrepancies between the survey and administrative records at the micro level. Key findings from Roemer's analysis include

- Distributions of DER wages for the two surveys were very similar, implying that differential sample selection bias was not a factor in SIPP's lower wage and salary income.
- Compared with the distribution of wages in the DER, SIPP had too many individuals with amounts below $30,000 and too few with amounts above $35,000; above $175,000, SIPP had only one-third to one-half as many earners as the DER.
- For 1996, the CPS had too few individuals with wages below $10,000, too many between $15,000 and $100,000, slightly too few between $100,000 and $200,000, and too many above $300,000.

[4]The increased reporting of Social Security benefits in the CPS lagged by a year the introduction of computer-assisted interviewing; nevertheless, the sudden stepped-up reporting suggests an instrument change.

- For sample members with both survey and DER wages, 57 percent of SIPP respondents and 49 percent of CPS respondents reported wages below their DER amounts; 3 percent of SIPP and 8 percent of CPS respondents reported wages equal to their DER amounts; and 40 percent of SIPP and 43 percent of CPS respondents reported wages above their DER amounts.
- The CPS appears to be superior to SIPP in capturing wages from the underground economy; in 1996, 3.6 percent of CPS wages and 1.8 percent of SIPP wages were reported by persons with no DER wages and no indication of self-employment; for the CPS this fraction grew from 2.5 percent in 1993.
- The CPS also appears to pick up more self-employment income misclassified as wages; in 1996, 3.0 percent of CPS wages and 1.5 percent of SIPP wages were reported by persons with no DER wages but with DER self-employment income; for the CPS this fraction grew from 2.2 percent in 1993.
- Both types of non-DER wages (underground wages and misclassified self-employment) occur at all income levels in both surveys, but the CPS has far more persons than SIPP with non-DER wages at upper income levels.

Thus, most of the difference between the SIPP and CPS wage and salary aggregates appears to be due to underreporting of legitimate wage income in SIPP, with misclassified self-employment income and the CPS's greater reporting of underground income accounting for less than a third of the gap between the two surveys.

Speculation about possible reasons for SIPP's underreporting of wage and salary income has focused on the possibility that the short reference period may lead SIPP respondents to report their take-home rather than gross pay despite the specificity of the questions. The short reference period, which is clearly helpful in capturing earnings from people with irregular employment, may also contribute to omissions of earned income. Roemer (2002) notes that when SIPP asked annual income questions at the end of each year, Coder (1988) found that the 12 months of reported wages for respondents with a single employer totaled nearly 7 percent less than what the same respondents reported in the annual round-up.

Income by Quintile

For most SIPP users, the quality of the income data in the lower end of the income distribution is far more important than its quality across the entire distribution. Furthermore, estimates of aggregate income for many sources are affected disproportionately by the amount of income captured in the upper tail of the distribution, in which the income holdings for those sources are concentrated. SIPP's superior capture of transfer income could reflect the survey's more complete capture of income in the lower end of the distribution generally.

To show how SIPP and CPS income estimates compare in different parts of the income distribution, Table A-4 presents estimates of aggregate income, by source, for quintiles of the population based on total family income, prepared for the panel.[5] Estimates are presented for 3 calendar years: 1993, 1997, and 2002. The SIPP estimates are from the 1992, 1996, and 2001 panels and, for consistency, are derived from the second year of data in each panel.[6] The CPS estimates are from the 1994, 1998, and 2003 supplements. The CPS data for all 3 years were collected with a computer-assisted instrument, whereas the SIPP data for 1993 were collected with a paper and pencil instrument. SIPP data for 2002 were the latest full calendar year available at the time the estimates in Table A-4 were prepared. By including comparative estimates for 2002, one can determine if the CPS gains during the first half of the 1990s persisted or whether the second new panel following the SIPP redesign was able to reverse the earlier trend. Unlike Roemer's estimates in Table A-1, the estimates in Table A-4 are based on public-use microdata files rather than the Census Bureau's internal files, and the 1993 SIPP estimates are from the second year of the 1992 panel rather than the first year of the 1993 panel. Also, the SIPP estimates in Table A-4 were calculated with the same method of aggregation used by Coder and Scoon-Rogers (1996), which differs from the method used by Roemer (2000) and Vaughan (1993). Differences between the percentages in the total column for 1993 and those reported in Table A-3 for comparable sources are due to any or all of these factors. Nevertheless, while there are differences by source, our estimate of SIPP aggregate income as a percentage of CPS aggregate income, at 94.5 percent, compares closely to Roemer's estimate of 94.8 percent.

The question of what happened to the ratio of SIPP to CPS income between 1997 and 2002 is answered by the estimates in the total column. While the ratio of SIPP to CPS total income declined from 94.5 to 89.0 percent between 1993 and 1997, the ratio rose slightly, to 89.4 percent, between 1997 and 2002. SIPP wages and salaries declined from 84.6 to 82.4 percent of the CPS aggregate, but this was offset by small improvements in every other source. On the whole, then, the relationships between income aggre-

---

[5]The bottom or first quintile contains the 20 percent of persons with the lowest family incomes. The top or fifth quintile contains the 20 percent of persons with the highest family incomes.

[6]The 1996 panel started 2 months late and did not collect data for all 12 months of 1996 for two of the four rotation groups.

gates in the two surveys appear to have stabilized following the movement that occurred with the introduction of computer-assisted interviewing in the CPS and the redesign of SIPP.

If one excludes the top quintile in order to eliminate the impact of differential topcoding as well as the CPS's seemingly more effective capture of very high incomes, one finds that the ratio of SIPP to CPS aggregate income increases by 4 to 6 percentage points in every year. SIPP wages and salaries and property income remain well below their CPS counterparts, but their shares of CPS income increase in all years. SIPP self-employment income remains well above the corresponding CPS amount, but the margin declines. For all other sources, the differences in their shares change little or in an inconsistent way when the top income quintile is excluded.

Turning to the results by income quintile, one finds, first, that the ratio of SIPP to CPS total income declines progressively from the bottom to the top quintile and does so in every year. Second, in the bottom quintile but no other quintile, the SIPP estimate of aggregate income exceeds the CPS aggregate in every year. Third, also in the bottom quintile alone, the ratio of SIPP to CPS income declines by as much between 1997 and 2002 as it did between 1993 and 1997, dropping from 119.5 to 112.2 percent and then to 105.7 percent of the CPS aggregate. In other words, over a period of only 9 years, SIPP went from capturing 20 percent more income than the CPS in the bottom quintile to capturing only 6 percent more income than the CPS. The 20 percent more income in 1993 included 25 percent more wages and salaries, 157 percent more self-employment income, 22 percent more property income, 7 percent more Social Security and Railroad Retirement income, 12 percent more Supplemental Security Income (SSI), an equal amount of welfare income, 24 percent more income from other transfers, and 44 percent more pension income. By 2002, SIPP was capturing only 9 percent more wages and salaries, 129 percent more self-employment income, 5 percent more property income, 12 percent *less* Social Security and Railroad Retirement income, 27 percent more SSI (an increase), 20 percent more welfare income (also an increase), 31 percent *less* income from other transfers, and 98 percent more pension income.

In the second income quintile, the SIPP captured 1.5 percent more aggregate income than the CPS in 1993, but this dropped to a 4 percent deficit by 1997. Unlike the first quintile, however, the SIPP held ground after that, gaining back a percentage point by 2002. The SIPP estimate of wages and salaries dropped from 100 percent of the CPS amount to 92 percent in 1997 but rose to 94 percent in 2002. Property income fell from 112 to 90 percent of the CPS amount, while Social Security and Railroad Retirement fell from 97 to 90 percent. Other transfers dropped from 90 to 59 percent of the CPS amount. Sizable improvements relative to the CPS were recorded for self-employment, SSI, welfare, and pensions, however.

TABLE A-4 SIPP Aggregate Income as a Percentage of CPS Aggregate Income by Source of Income and Family Income Quintile: 1993, 1997, and 2002

	Quintile of Family Income						
Income Source	Bottom	Second	Third	Fourth	Top	Total	Bottom Four
				1993 Calendar Year			
Total Income	119.5	101.5	95.9	93.0	88.9	94.5	98.2
Wages and salaries	124.8	100.1	92.6	88.7	81.8	88.9	94.0
Self-employment	257.4	133.3	128.8	139.3	178.0	160.3	139.0
Property income	121.7	112.4	101.5	99.0	67.3	83.5	104.2
Social Security and Railroad Retirement	107.2	97.4	98.6	105.7	107.4	102.2	101.8
Supplemental Security Income	111.6	77.8	75.6	89.5	91.5	98.2	98.5
Welfare	99.0	72.4	114.6	180.2	141.0	97.6	97.0
Other transfers	123.5	89.5	84.6	81.2	72.9	87.9	92.3
Pensions	143.6	116.9	110.6	115.9	107.3	113.9	116.3
				1997 Calendar Year			
Total Income	112.2	96.0	93.8	91.8	80.7	89.0	95.3
Wages and salaries	111.7	91.7	89.7	87.9	77.2	84.6	90.6
Self-employment	214.7	152.2	153.0	160.5	169.3	165.6	159.5
Property income	112.6	90.6	74.7	64.2	32.4	49.0	75.9
Social Security and Railroad Retirement	100.0	89.3	97.4	99.1	75.5	93.6	95.7
Supplemental Security Income	122.0	126.1	121.7	166.7	149.0	126.5	125.7
Welfare	122.4	143.5	267.2	420.1	509.9	145.0	140.4
Other transfers	71.6	58.7	57.7	52.5	41.4	54.7	59.3
Pensions	229.0	150.5	136.9	142.9	102.6	134.5	148.6

2002 Calendar Year

Total Income	105.7	97.3	92.8	90.7	83.0	89.4	94.2
Wages and salaries	109.4	93.8	86.0	85.3	75.2	82.4	88.6
Self-employment	219.5	162.4	160.5	158.1	208.6	188.3	163.2
Property income	104.7	90.0	70.7	56.0	38.6	52.3	70.5
Social Security and Railroad Retirement	87.6	90.4	104.4	115.9	94.5	95.4	95.4
Supplemental Security Income	126.6	121.3	139.4	152.6	222.9	131.5	128.5
Welfare	119.5	109.4	176.4	246.6	1,415.5	146.7	128.4
Other transfers	70.9	58.8	64.2	57.4	42.8	58.0	62.4
Pensions	198.4	140.2	155.7	153.5	128.3	147.0	153.9

SOURCE: 1992, 1996, and 2001 SIPP panels and 1994, 1998, and 2003 CPS ASEC supplements.

These basic patterns were repeated in the third and fourth quintiles, for which the SIPP estimates of aggregate income fell by 2 to 3 percentage points relative to the CPS, ending up at 93 percent in the third quintile and 91 percent in the fourth quintile. There was one notable exception to the patterns by income source; the capture of Social Security and Railroad Retirement in SIPP improved between 1997 and 2002 to the point at which SIPP captured relatively more of such income in comparison to the CPS in 2002 than in 1993. In both cases, the SIPP aggregates in 2002 exceeded the CPS aggregates. Elsewhere, SIPP fell further behind the CPS in wages and salaries, property income, and other transfers but improved in self-employment, SSI, welfare, and pensions. In both quintiles, SIPP captured 50 percent more pension income in 2002 than did the CPS.

In the top quintile, the SIPP estimate of aggregate Social Security and Railroad Retirement dropped relative to the CPS, as it did in the first and second quintiles, and the relative gain in the capture of pension income was more modest than in the lower quintiles. Otherwise, the different sources improved or declined in the SIPP, just as they did in the lower quintiles. Over all sources, the SIPP estimate of total income in the top quintile dropped from 89 percent of the CPS estimate in 1993 to 81 percent in 1997 but then rose to 83 percent in 2002.

On balance, then, while there was gradual erosion in the amount of income collected by SIPP relative to the CPS between 1990 and 1996, which was due largely to changes in the CPS, there was a much more substantial reduction in the relative amount of income that SIPP collected from the bottom quintile of the family income distribution. This change is significant because it detracts from what has been SIPP's greatest strength, historically, in the collection of income data.

PROGRAM PARTICIPATION

SIPP was designed to do a better job of capturing program participation and benefit amounts than other surveys, and from the beginning it has generally done so. SIPP still falls short of administrative totals for most programs, but in some cases SIPP estimates exceed the program totals.

A recent review of survey reporting of program participation and benefit receipt relative to administrative totals concludes that SIPP "typically has the highest reporting rate for government transfers, followed by the CPS" and the Panel Study of Income Dynamics (PSID), but that some programs—specifically unemployment insurance and workers' compensation—are reported more fully in the CPS than in SIPP (Meyer, Mok, and Sullivan, 2009). The study also finds the highest overall dollar reporting in SIPP and the American Community Survey (ACS). These are followed, in turn, by the CPS, PSID, and the Consumer Expenditure (CE) Survey. One other conclusion of note is that while the reporting of most programs in the PSID, CPS, and CE experienced a significant decline over time, the decline was less pronounced in SIPP and the ACS actually showed improvement.

Table A-5 presents shares of administrative totals of average monthly participants and aggregate annual benefits estimated by SIPP and the CPS in 1987, 1996, and 2005 for the following programs:

- Food Stamp Program (FSP)
- Aid to Families with Dependent Children (AFDC)/Temporary Assistance for Needy Families (TANF)
- Old-Age and Survivors Insurance (OASI)
- Social Security Disability Insurance (SSDI)
- Supplemental Security Income (SSI)

TABLE A-5 SIPP and CPS Estimates of Program Participants and Aggregate Benefits as a Percentage of Administrative Benchmarks, Selected Years

	1987		1996		2005	
Program	SIPP	CPS	SIPP	CPS	SIPP	CPS
	Survey Estimate of Average Monthly Participants as a Percentage of Administrative Benchmark					
FSP	88.1	73.2	84.2	66.3	82.9	56.5
AFDC/TANF	76.4	80.5	79.5	67.0	80.9	63.0
OASI	94.5	88.0	94.2	84.3	97.1	82.9
SSDI	101.3	101.6	90.6	89.7	93.8	78.8
SSI	90.3	72.7	94.4	65.8	102.7	58.0
NSLP	113.7	54.4	111.6	64.2	112.5	48.4
WIC	66.0		56.1		58.2	60.1
	Survey Estimate of Aggregate Benefits as a Percentage of Administrative Benchmark					
FSP	85.9	74.2	79.0	63.1	76.4	54.6
AFDC/TANF	73.0	74.4	77.0	66.8	62.2	48.7
OASI	95.0	89.0	88.7	90.5	97.4	89.7
SSDI	95.4	100.1	79.3	91.9	84.8	81.8
SSI	89.6	76.7	93.4	77.6	110.0	79.4

NOTE: FSP = Food Stamp Program; AFDC/TANF = Aid to Families with Dependent Children/ Temporary Assistance for Needy Families; OASI = Old-Age and Survivors Insurance (Social Security); SSDI = Social Security Disability Insurance; SSI = Supplemental Security Income; NSLP = National School Lunch Program; WIC = Special Supplemental Nutrition Program for Women, Infants and Children.

SOURCE: Adapted from Meyer, Mok, and Sullivan (2009).

- National School Lunch Program (NSLP)
- Special Supplemental Nutrition Program for Women, Infants and Children (WIC)

For every program but WIC, SIPP captures a substantially higher percentage of average monthly participants than the CPS in 2005, but was no better than the CPS for AFDC in 1987 and SSDI in both 1987 and 1996. Benefits, which are reported about as well as participants for some programs but less well for others, tell a similar story. In 2005, SIPP captured a higher share of aggregate annual benefits than the CPS for the FSP, AFDC/TANF, OASI, and SSI, but was only marginally better for SSDI. In 1987, SIPP was on a par with the CPS for AFDC/TANF and SSDI, which is consistent with the surveys' relative estimates of participants in these two programs in that year.

Whether because of poor recall or because respondents sometimes answer on the basis of their current situation, CPS estimates of persons who ever participated in a program sometimes line up with SIPP estimates of average monthly participants. Medicaid provides a good example. SIPP's estimate of Medicaid participants in a given month is comparable to the CPS's estimate of persons who were ever covered by Medicaid during the previous year. For instance, SIPP estimated that 11.8 percent of the population was covered by Medicaid in December 2002, whereas the CPS estimated that 11.6 percent of the population was ever enrolled in Medicaid in the 2002 calendar year (Czajka and Denmead, 2008). SIPP's estimate of persons ever enrolled in Medicaid was 17.1 percent of the population—substantially larger than the CPS estimate.

INCOME RECEIPT FROM MULTIPLE SOURCES

One of the early uses of SIPP data was to support estimates of multiple benefit receipt by participants in transfer programs and to help determine how often the receipt of benefits from more than one program was serial versus simultaneous (see, for example, Doyle and Long, 1988). High-quality monthly data were critical to such research, and SIPP has remained unique in its ability to support the production of such estimates (see, more recently, Reese, 2007).

A related area of research involves determining the extent to which the beneficiaries of a given program are dependent on that program for their economic support. For example, the fraction of total income that retired persons derive from Social Security benefits is highly relevant to policy debates involving how to ensure the continued financial solvency of the Social Security system. Estimates from the CPS for calendar year 2001 show that 22 percent of retired workers relied on their Social Security payments for 100 percent of their income (Czajka, Mabli, and Cody, 2008). The corresponding figure from SIPP was only 8 percent.[7] Similarly, the CPS finds 18 percent of retired workers receiving less than 25 percent of their income from Social Security payments, and SIPP finds nearly twice that share, or 30 percent. Retired workers in SIPP were more likely than their counterparts in the CPS to report receiving each of six additional sources: wages, self-employment, property income, pensions, SSI, and welfare.[8] These differences highlight the impact of SIPP's short reference period and the survey's focus on income and benefit recipiency.

WEALTH

One of the key limitations of the CPS for modeling program eligibility is the absence of data on assets and liabilities. Many programs include in their eligibility criteria some limitations on asset holdings, and some programs have very explicit provisions about particular types of assets, such as vehicles. From its inception, SIPP has collected data on asset holdings and liabilities, and its focus on the types of assets and debts held by low-income families has been an important feature of the SIPP wealth data.

The standard against which all survey data on wealth are measured is the Survey of Consumer Finances (SCF), conducted by the Federal Reserve Board. In addition to hundreds of questions on detailed components of assets and liabilities, the SCF includes a high-income subsample drawn from tax records. Wealth is even more heavily concentrated than income, with about a third of all wealth held by the wealthiest 1 percent of families and two-thirds held by the wealthiest 10 percent, leaving one-third for the remaining 90 percent of the population (Kennickell, 2006). Accurate measurement of aggregate wealth holdings requires a sample design that reflects this distribution.

Wolff (1999) compared SIPP, the SCF, and the PSID with respect to a number of measures of the size and distribution of wealth over the mid-1980s through the mid-1990s. His findings suggest that, for the lowest two income quintiles, SIPP did as well as the SCF in capturing asset holdings. Furthermore, SIPP's comparative performance did not deteriorate a great

[7]Fisher (2007) reports that for 1996 the survey estimates of full reliance on Social Security benefits were 17.9 percent for the CPS and 8.5 percent for SIPP, so the difference between the two surveys appears to have grown between 1996 and 2001. The 1996 difference was reduced only slightly when matched administrative records were substituted for survey data and used to assign beneficiary status.

[8]Earlier, we documented that underreporting of SSI and AFDC/TANF is greater in the CPS than in SIPP. Matches to program administrative records show that while Social Security beneficiaries in both surveys nearly always report receiving benefits, SSI beneficiaries are significantly less likely to report receiving SSI in the CPS than in SIPP (Koenig, 2003).

deal through the next two quintiles—that is, through the lower 80 percent of the income distribution. SIPP also did particularly well in capturing the major *types* of wealth held by the middle class, such as homes, vehicles, and savings bonds, but it did not do as well in capturing the types of assets held by the wealthiest families.

A comparison of the three surveys' estimates of wealth in late 1998 and early 1999 showed that SIPP's estimate of aggregate net worth, defined as assets minus liabilities, of $14.4 trillion was just under half of the SCF estimate of $29.1 trillion and 60 percent of the PSID estimate (Czajka, Jacobson, and Cody, 2003). The SIPP estimate of median net worth, $48,000, was two-thirds of the SCF median of $71,800 and 74 percent of the PSID median.

SIPP is much more effective in capturing liabilities than assets. SIPP's estimate of aggregate assets was 55 percent of the SCF estimate of $34.1 trillion, but its estimate of aggregate liabilities was 90 percent of the SCF estimate of $5.0 trillion. SIPP's estimate of median assets was 83 percent of the SCF median of $116,500, while its estimate of median liabilities was 97 percent of the SCF median of $11,900. SIPP's weaker performance in measuring net worth than either assets or liabilities reflects the imbalance in the survey's estimates of these two components. By estimating the negative side of the balance sheet more fully than the positive side, SIPP adds to its underestimate of net worth.

As a proportion of the corresponding SCF estimate, SIPP's estimates of aggregate assets exhibit wide variation by type. SIPP's estimate of the value of the home was 91 percent of the SCF estimate, but SIPP captured only 41 percent of the SCF valuation of other real estate. SIPP captured 76 percent of the SCF estimate of motor vehicles but only 17 percent of SCF business equity. Among financial assets, SIPP's estimate of 401(k) and thrift accounts was 99 percent of the SCF estimate, but the next best component, other financial assets, was only 71 percent of the SCF estimate. For assets held at financial institutions, the SIPP estimate was 63 percent of the SCF estimate. For stocks and mutual funds, the largest financial asset, the SIPP estimate was only 59 percent of the SCF estimate, whereas the SIPP estimate of IRA and Keogh accounts was 55 percent of the SCF estimate. Finally, the SIPP estimate of other interest earning assets was only 33 percent of the SCF amount.

In contrast to Wolff's findings from the mid-1980s and early 1990s that SIPP matched the SCF in capturing the asset holdings of the bottom two income quintiles, the estimates from the 1996 panel showed that SIPP did not fare appreciably better with the assets of low-income families than with higher income families. Between the early and late 1990s, SIPP families with negative or zero net worth grew from 13 to 17 percent of the population while SCF families with no net worth remained at 13 percent.

A more telling sign of the reduction in the quality of the SIPP wealth data is that the correlation between assets and liabilities dropped from .49 in the 1992 and 1993 SIPP panels to between .06 and .19 in the 1996 panel. Over the same period the correlation in the SCF dropped only moderately, from .50 to .40.

HEALTH INSURANCE COVERAGE TRANSITIONS

Average monthly estimates of health insurance coverage from SIPP compare closely to estimates of health insurance coverage obtained in the National Health Interview Survey (NHIS), which measures coverage at the time of the interview and therefore is free of recall bias (Czajka and Denmead, 2008; Davern et al., 2007). Both sets of estimates are also relatively close to the estimate of persons with health insurance coverage from the CPS. The CPS estimate is intended to measure any amount of coverage over the prior calendar year and therefore ought to be higher than both the SIPP and NHIS estimates, but it clearly suffers from some combination of underreporting and a tendency for respondents to answer in terms of their current coverage. To its strong cross-sectional estimates of health insurance coverage, SIPP adds a longitudinal dimension, which enables the survey to provide information on changes in coverage over time and what coverage people ever had over an extended period of time.

One of the objectives reflected in the design of SIPP is to obtain more reliable data on short-term dynamics by interviewing respondents three times a year and asking them to recall events as recently as the prior month and no more than 4 months earlier. While the potential benefits of a shortened reference period are obvious, frequent interviews also create more opportunities for erroneous reports. For example, respondents who are uncertain about the type of health insurance coverage they have may give different responses in different waves. In addition, a sample member may self-report in one wave but have his or her data reported by a proxy respondent in the next wave. A misreported status in one wave creates two false transitions, compounding the effects of a single error.

Lacking a good benchmark for assessing the reliability of reported transitions, researchers have produced very little evidence regarding the quality of transitions in health insurance coverage in the SIPP. It is true, for example, that SIPP finds a higher proportion of the population who were ever without health insurance coverage during a 12-month period than does the NHIS, which relies on a retrospective question (Czajka and Denmead, 2008). Does this suggest that SIPP may be overestimating periods without coverage, or does the difference between the two surveys simply reflect SIPP's better design for estimating incidence over time? The shorter recall required in SIPP suggests the latter, but SIPP also finds a higher proportion of people ever uninsured during a year than the longitudinal Medical Expenditure Panel Survey (MEPS), which, like SIPP, conducts multiple interviews over a year. MEPS differs from SIPP in having a variable reference period. As yet, there is no definitive answer to the question of which survey is more correct. However, the high rate at which SIPP sample members—particularly children—transition between the uninsured and insured does raise questions about the SIPP estimates. In the 2001 SIPP panel, 44 percent of uninsured children gained coverage between one wave and the next. This compares with 23 percent for uninsured adults ages 19 to 39 and 19 percent for uninsured adults ages 40 to 59 (Czajka and Mabli, 2009).

A review of coverage transitions measured in SIPP found instances of improbable transitions that appeared likely to be reporting errors—for example, children losing and regaining employer-sponsored coverage through a parent who reported continuous coverage over the same period. Edits to remove improbable transitions such as these reduced the estimated number of one-wave uninsured spells among children by 52 percent (Czajka and Sykes, 2006). The reductions among adults were smaller: 31 percent for adults ages 19 to 39 at the start of the wave and 22 percent among adults ages 40 to 64. Clearly, the frequency of brief uninsured spells and reported transitions into and out of the uninsured status in SIPP should be a matter of concern among users. However, data sources that can provide accurate reports of monthly status and be linked to the SIPP are few in number, and only the Census Bureau is legally able to produce the linkages that will support such research.[9]

ATTRITION

In addition to nonresponse at the initial interview, which has tended to be quite low in comparison to other household surveys, SIPP as a longitudinal survey is subject to attrition of sample members over time. The bias that attrition may introduce into survey estimates makes the level of attrition a serious concern.

Sample Loss

Table A-6 documents both incremental and cumulative sample loss due to nonresponse by wave in each of the four SIPP panels that started

[9]Such data sources include Medicaid administrative files from the Medicaid Statistical Information System, which have been linked to the CPS and NHIS but not SIPP, and Internal Revenue Service Forms 5500 filed by employers and processed by the U.S. Department of Labor's Employee Benefit Security Agency (regarding the latter, see Decressin, Hill, and Lane, 2006).

TABLE A-6 Incremental and Cumulative Household Sample Loss Rates by Wave: 1992, 1993, 1996, and 2001 SIPP Panels, Unweighted (percentage)

	Incremental Sample Loss Rate				Cumulative Sample Loss Rate			
Wave	1992 Panel	1993 Panel	1996 Panel	2001 Panel	1992 Panel	1993 Panel	1996 Panel	2001 Panel
1	9.3	8.9	8.4	13.3	9.3	8.9	8.4	13.3
2	5.3	5.3	6.1	8.6	14.6	14.2	14.5	21.9
3	1.8	2.0	3.3	2.8	16.4	16.2	17.8	24.7
4	1.6	2.0	3.1	1.2	18.0	18.2	20.9	25.9
5	2.3	2.0	3.7	1.6	20.3	20.2	24.6	27.5
6	1.3	2.0	2.8	0.7	21.6	22.2	27.4	28.2
7	1.4	2.1	2.5	0.7	23.0	24.3	29.9	28.9
8	1.7	1.2	1.4	1.4	24.7	25.5	31.3	30.3
9	1.5	1.4	1.5	1.6	26.2	26.9	32.8	31.9
10	0.4		1.2		26.6		34.0	
11			1.1				35.1	
12			0.4				35.5	

NOTE: The household sample loss rate expresses the number of noninterviews among eligible households in a given wave as a percentage of the total eligible households in that wave. Eligible households include those that the Census Bureau continues to attempt to interview as well as those that have been dropped from further interview attempts in keeping with SIPP field procedures but remain within the SIPP universe. Households dropped from further interview attempts include nonrespondents to the Wave 1 interview as well as households that were interviewed in Wave 1 but missed two or three consecutive interviews (depending on the reason) or moved too far from a SIPP primary sampling unit. All noninterviewed households (except those known to have left the survey universe) are multiplied by a growth factor to reflect a crude estimate of households splitting to form multiple households less those leaving the SIPP universe. Beginning with Wave 4 of the 2001 panel, households are no longer dropped from further interview attempts because they missed consecutive interviews.
SOURCE: Eargle (2004).

between 1992 and 2001. The estimates of sample loss apply to eligible households. If at least one member of an eligible household responds during a given wave, the Census Bureau collects or imputes data for every other household member. An eligible household contributes to the estimate of sample loss in a given wave if no interview is conducted with any member of that household.

While initial nonresponse declined slightly over the 1992, 1993, and 1996 panels, it jumped nearly 5 percentage points between the 1996 and 2001 panels, rising from 8.4 to 13.3 percent.[10] This increase in household

[10]An incentive experiment that paid $10 or $20 to about half the Wave 1 sample households in the 1996 panel contributed to the low sample loss in Wave 1 and subsequent waves (see James, 1997).

nonresponse did not begin with the 2001 panel, however. The incremental sample loss rate for every wave after the first rose between the 1993 and 1996 panels. At the end of Wave 9, the cumulative sample loss rate for the 1996 panel stood at 32.8 percent versus 26.9 percent in the 1993 panel. The 1996 panel ran three additional waves, but the cumulative sample loss grew by less than 3 percentage points—to 35.5 percent—over those three waves.

For comparison purposes, Table A-7 reports nonresponse rates to the CPS ASEC supplement and the labor force survey conducted in the same month.[11] Some households that complete the monthly labor force survey do not respond to the supplement. Historically, nonresponse to the monthly labor force survey has been very low. Noninterview rates deviated little from 4 to 5 percent of eligible households between 1960 and 1994 but then began a gradual rise that coincided with the introduction of a redesigned survey instrument using computer-assisted interviewing (U.S. Census Bureau, 2002). By March 1997, the first data point in Table A-7, the noninterview rate had reached 7 percent, but it rose by just another percentage point over the next 7 years. Over this same period, nonresponse to the ASEC supplement among respondents to the labor force survey ranged between 8 and 9 percent, with no distinct trend, yielding a combined sample loss that varied between 14 and 16 percent of the eligible households. In other words, the initial nonresponse to the 2001 SIPP panel is still 2 to 3 percentage points lower than the nonresponse to the ASEC supplement. But as a measure of how much the SIPP response rates have declined, it took *two* waves of cumulative sample loss in the 1996 panel to match the nonresponse to the ASEC supplement.

A SIPP practice dating back to the start of the survey bears some responsibility for the amount of sample loss after Wave 3 in panels prior to 2001. Households that missed two or three consecutive interviews (depending on the circumstances) were dropped from further attempts. The principal purpose, initially, was to ensure that all missing waves would be bounded by complete waves, so that the missing waves could be imputed from the information collected in the surrounding waves. Missing wave imputations were performed for the first time in the early 1990s but were discontinued with the 1996 redesign. With rising attrition and the removal of the principal rationale for dropping respondents after two missing waves, the Census Bureau revised this practice during the 2001 panel. Respondents are no longer dropped after missing two consecutive interviews. The impact

[11]Until 2001 the CPS supplement that collects annual income was conducted solely in March of each year, but as part of a significant sample expansion, the Census Bureau began to administer the supplement to CPS sample households in February and April that were not interviewed in March.

TABLE A-7 Nonresponse to the CPS Labor Force Survey and ASEC Supplement, 1997 to 2004

Sample Year	Percentage of Eligible Households Not Responding to the Labor Force Questionnaire	Percentage of Labor Force Respondents Not Responding to the Supplement	Percentage of All Eligible Households Not Responding to the Supplement
1997	7.2	9.2	15.7
1998	7.8	7.2	14.4
1999	7.9	8.9	16.1
2000	7.0	8.0	14.4
2001	8.0	8.5	15.9
2002	8.3	8.6	16.2
2003	7.7	8.0	15.0
2004	8.5	8.2	16.0

NOTE: March 1997 is the first supplement for which the CPS technical documentation reports rates of nonresponse. The nonresponse rate in column 3 is the sum of the nonresponse rate in column 1 and the product of the nonresponse rate in column 2 (divided by 100) and 100 minus the nonresponse rate in column 1.
SOURCE: Current Population Survey Technical Documentation, various years.

of the new policy is evident in the incremental sample loss rate between Waves 3 and 4, which dropped to 1.2 percent from a level of 3.1 percent in the 1996 panel. By Wave 7 the cumulative sample loss had fallen below that of the 1996 panel, which meant that the survey had retained enough additional sample members to offset both the 5 percentage point higher Wave 1 nonresponse rate and higher attrition between Waves 1 and 2. The 2001 panel maintained a lower cumulative sample loss through the remaining two waves. Interestingly, the incremental sample loss rates between Waves 8 and 9 were essentially identical across the four panels at about 1.5 percent.

Attrition Bias

Numerous studies with SIPP and other panel surveys have documented that attriters differ from continuers in a number of ways (see, for example, Fitzgerald, Gottschalk, and Moffitt, 1998; Zabel, 1998). Most studies of attrition bias have been limited to comparing attriters and continuers with respect to characteristics measured at the beginning of the survey, before any attrition has occurred. Such studies cannot say how much the attriters and continuers differ on characteristics subsequent to attrition, which is critical to knowing how longitudinal analyses may be affected by attrition. Another limitation of such studies that is rarely noted is that they assume that the quality of the data provided by those who will later leave is comparable to that provided by those who remain in the panel through its conclusion. For many characteristics, this assumption is probably valid. But for sensitive characteristics or those that respondents might view as onerous to provide, the validity of the assumption is questionable. Yet another limitation of many attrition studies is that they fail to separate nonrespondents who left the survey universe from those who remained eligible to be interviewed. Persons who leave the survey universe—by dying, joining the military, becoming institutionalized (including incarceration), or moving outside the country—have distinctly different characteristics than those who remain in the universe.

Administrative records linked to survey data can overcome these limitations. Administrative records can provide data on postattrition and even presurvey characteristics, and the values of the characteristics are recorded with very little error, generally. Moreover, any measurement error in the characteristics obtained from administrative records will be independent of attrition status. Finally, most nonrespondents who left the survey universe are identified in SIPP and can be removed from the sample of attriters. Some who cannot be identified in the survey data may drop out of analyses automatically because their administrative records terminate at some point after they have left the survey universe.

Vaughan and Scheuren (2002) used Social Security Administration Summary Earnings Records matched to SIPP panel data to compare attriters and continuers with respect to earnings and program benefits over time.[12] Even after removing those who left the survey universe, they found that attriters and nonattriters differed markedly with respect to earnings and receipt of program benefits at the beginning of a panel—that is, before any attrition had occurred. Over time, however, these differences attenuated. With enough passing years (longer than a typical SIPP panel, however), the characteristics of those who left and those who continued to respond to the survey converged. This trend suggests that compensating for the impact of attrition on cross-sectional estimates becomes both easier and less important over time. But the fact that the differences are large to begin with and then diminish over time also implies that attriters experience greater change than nonattriters. Vaughan and Scheuren (2002) concluded that compensating for the attrition bias in estimates of gross change is both important and much more difficult than compensating for differences in net change.

To evaluate the effectiveness of the Census Bureau's nonresponse adjustments, Czajka, Mabli, and Cody (2008) used administrative data

[12]Vaughan and Scheuren (2002) examined attrition in the Survey of Program Dynamics, which was selected from the 1992 and 1993 SIPP panels, and continued to interview respondents through 2002.

from the same sources as Vaughan and Scheuren but compared the full sample—using a Wave 1 cross-sectional weight—with the subsample of continuers weighted by the full panel weight, which incorporates adjustments for differential nonresponse.[13] They found little evidence of bias in estimates of earnings, Social Security beneficiary status and benefit amounts, or SSI beneficiary status and benefit amounts at different points in time. Nor did they find significant bias in selected estimates of change in these characteristics. The implication is that attrition bias in these characteristics is being addressed in the longitudinal weights. It is not possible to evaluate the Census Bureau's adjustments to the cross-sectional weights in the same manner as the longitudinal weights, as there is no attrition-free cross-sectional sample after the first wave. Furthermore, other, lesser known biases due to attrition are not addressed by the weights. For example, Czajka and Sykes (2006) documented attrition bias among new mothers, which contributes to a severe underestimate of the number of infants if the weights of mothers are assigned to their newborn children.[14] Attrition by new mothers has been documented in the National Longitudinal Survey of Youth 1997 as well, although, in that survey, becoming a parent was found to be very highly related to returning to the survey after missing an interview (Aughinbaugh and Gardecki, 2007).

REPRESENTATION OF THE POPULATION OVER TIME

Although SIPP is fundamentally a panel survey, cross-sectional applications (including analysis of repeated cross-sections) abound and may in fact be more common than true longitudinal uses of the data. For this reason, it is important that users understand the limits to the survey's representation of the population over time.

While the U.S. population is currently growing at a rate of less than 1 percent a year, this net growth is the difference between substantially larger inflows and outflows. SIPP panel members who leave the sample by dying, entering institutions, moving abroad, or moving into military barracks represent the outflows from the population. A priori, there is no reason to think that SIPP underrepresents, overrepresents, or otherwise misrepresents the gross outflows from the population, although one could certainly speculate that respondents who know that they are moving abroad or entering institutions may leave before being identified as leaving the survey universe.

---

[13]Persons who leave the SIPP universe are assigned panel weights if they missed no prior interviews. Such persons will have contributed to both the full sample and the panel estimates.

[14]SIPP longitudinal weights are not assigned to persons entering the sample after the calendar month to which the weights are calibrated. It is common among users to assign infants the weights of their mothers.

To maintain full cross-sectional representativeness over time, however, a panel survey must also obtain—periodically if not continuously—a representative sample of new entrants to the population. New entrants include births, immigrants, and persons returning from abroad. Because SIPP excludes residents of specific types of group quarters (prisons, nursing homes, and military barracks, primarily), new entrants also include persons moving from such quarters into households. SIPP captures births to panel members and, through this mechanism, represents most births to the population over the length of a panel, but its capture of other new entrants is limited to persons moving into households with original sample members. That is, SIPP represents those additional new entrants who join households containing persons who were in the SIPP universe at the start of a panel. SIPP does not represent people who enter or reenter the U.S. civilian noninstitutionalized population if they form new households or join households populated by people who have also joined the population since the start of the panel. What fraction of new entrants other than births is represented in SIPP is unknown and not readily discernible. New entrants are not identified explicitly in the SIPP public-use data files, and, even if they were, none of the SIPP weights is designed to properly reflect their contribution to the population. An estimate of the *total* new entrant population, exclusive of births, near the end of the 1996 SIPP panel placed it at about 10 million, or more than 3 percent of the total population (Czajka and Sykes, 2006). This estimate represents how many persons, other than those born to panel members, were in the civilian noninstitutionalized population at the end of the 1996 panel but had not been in the population at the start of the panel.

To facilitate cross-sectional uses of SIPP data, the Census Bureau provides monthly cross-sectional weights. These weights include an adjustment for differential attrition and a separate "mover adjustment," which offsets the weights assigned to persons who join SIPP households. In addition, the cross-sectional weights are poststratified to monthly estimates of the civilian noninstitutionalized population by age, gender, race, and Hispanic origin. This poststratification to demographic controls is a limited attempt to make the SIPP sample consistent with changes in the size and composition of the civilian noninstitutionalized population over time. Poststratification ensures that the monthly SIPP cross-sectional weights will sum to the Census Bureau's estimates of monthly population totals by age, gender, race, and Spanish origin. It does not ensure that the broader characteristics of the weighted sample will remain consistent with the population over time if the net effect of the gross inflows and outflows is to change the characteristics of the population.

The implications of these population flows for the representativeness of the SIPP cross-sectional sample over time is unknown, and the issue has attracted very little interest. But analysis of the characteristics of SIPP sample members who move out of the population over time indicates that these people differ dramatically from nonmovers with similar demographic characteristics (particularly those of Hispanic origin). This implies a potential for persons moving into the population to differ dramatically as well (Czajka, 2007). Within-panel trends that have been attributed to attrition could very well be owing to the panel's increasingly less complete representation of the national population over time as the new entrants omitted from the SIPP grow from zero to as much as 3 percent of the total population. If so, then a new strategy for weighting SIPP that takes account of the new entrants who are not represented by the survey could improve the quality of inferences supported by the data.[15]

SEAM BIAS

Seam bias describes a tendency for transitions to be reported at the seam between survey waves—that is, between month 4 of one wave and month 1 of the next wave—rather than within waves. Evidence of seam bias was first identified in analyses of the Income Survey Development Program research panels that preceded the SIPP (Callegaro, 2008). Multiple causes have been suggested, and the causes appear to be multiple in nature. The extent of seam bias varies markedly across items, which may reflect different mixes of causes. SIPP users have adapted their analytical strategies. It is common for those examining behavior over time to take only one data point per wave—either the one calendar month that is common to all four rotation groups or the fourth reference month, which is widely viewed as the most reliable because of its proximity to the interview month. The inference is that there is not enough independent information in the other three months to make them analytically useful or that analysts do not know how to use the limited additional information that they provide. The Census Bureau has tried two alternative approaches to dealing with seam bias: (1) collecting selected data for the interview month as a fifth reference month, which will overlap the first reference month of the next wave, and (2) dependent interviewing. It remains unclear what the Census Bureau has learned from collecting the additional month of data. These data are not

[15]If the survey with its current cross-sectional weights underestimates poverty in the full population, for example, because it underrepresents 10 million people with a very high poverty rate, then one strategy would be to exclude the 10 million from the weighted population total so that the poverty rate estimated from the survey provides a better reflection of the population to which the weights sum. An alternative strategy, if the characteristics of the 10 million can be known sufficiently well to be replicated within the existing survey sample, is to revise the cross-sectional weighting of the sample to better reflect the characteristics of the total population.

released on the public-use file, and it is not apparent that the Census Bureau has made use of this information in editing responses, which might have moved the seam by one month but not reduced it. However, the Census Bureau appears to have had some success with dependent interviewing, in which respondents who reported participation in a program at the end of the previous wave are informed of their prior wave response and asked if they were still participating 4 months earlier. Specifically, dependent interviewing has helped to lower the frequency of transitions at the seam by reducing the number of reported transitions rather than shifting their location (Moore et al., 2009). However, dependent interviewing has given rise to other problems during its application to the 2004 panel, and the Census Bureau has suspended its use in SIPP.

IMPUTATION

Item nonresponse is higher on income questions than on most other types of questions.[16] Since the start of SIPP, item nonresponse to income questions in surveys has increased dramatically. This is reflected in the proportion of total income that is imputed.

Growth of Imputation Over Time

In 1984, just 11.4 percent of total money income in SIPP was imputed (Vaughan, 1993). Even then, however, imputation rates varied widely across income sources. Income imputation was lowest for public assistance (7.5 percent) and highest for property income (23.9 percent). The single highest imputation rate occurred for dividends (46.8 percent), a component of property income. The imputation rate for wage and salary income was among the lowest at 8.8 percent. Imputation rates in the CPS were higher—in large part because the Census Bureau imputes the entire ASEC supplement for respondents who complete only the brief monthly labor force survey that precedes the supplement. In March 1985, 20.1 percent of total CPS ASEC income for 1984 was imputed—including 17.9 percent of wage and salary income.

Between 1984 and 1993, imputation rates for SIPP income increased substantially, growing to 20.8 percent for total income and 17.7 percent for wages and salaries, or double the rate in 1984 (see Table A-8). The imputation rate for property income, 42.4 percent, approached the very high level recorded by dividends in 1984. The low imputation rate for public assistance as a whole grew to more than 13 percent for SSI and welfare.

[16]Item nonresponse on asset questions is even higher.

TABLE A-8 Proportion of Income Imputed, by Source: SIPP and CPS, Selected Years

Income Source	Survey Reference Year		
	1993	1997	2002
		SIPP	
Total Income	20.8	24.0	28.6
Wages and salaries	17.7	20.5	24.9
Self-employment	29.3	32.7	36.4
Property income	42.4	42.9	49.7
Social Security and Railroad Retirement	22.6	22.7	28.8
Supplemental Security Income	13.2	16.4	22.6
Welfare income	13.8	31.2	32.8
Other transfers	20.8	33.0	33.6
Pensions	23.7	37.3	47.3
		CPS	
Total Income	23.8	27.8	34.2
Wages and salaries	21.5	24.8	32.0
Self-employment	34.6	39.5	44.7
Property income	42.4	52.8	62.6
Social Security and Railroad Retirement	24.1	27.9	35.5
Supplemental Security Income	22.9	19.7	28.0
Welfare income	19.8	18.1	29.2
Other transfers	23.3	23.9	31.4
Pensions	24.2	27.0	35.4

SOURCE: The 1992, 1996, and 2001 SIPP panels and the 1994, 1998, and 2003 CPS ASEC supplements.

Between 1993 and 2002, the proportion of total income that was imputed increased by 8 percentage points. The increase in imputation rates by income source was very uneven. The income imputation rates for welfare, other transfers, and pensions surged between 1993 and 1997. For welfare, the imputation rate more than doubled, rising from 14 to 31 percent. For other transfers and pensions, the imputation rates increased by more than half, reaching 33 percent for other transfers and 37 percent for pensions. Yet there was no increase in the already high imputation rate for property income, and the imputation rates for wages and salaries and self-employment income increased by only 3 percentage points. Between 1997 and 2002, the imputation rate for pension income grew another 10 percentage points, taking it very near the imputation rate for property income, which grew by 7 percentage points to nearly 50 percent. Imputation rates for both wages and salaries and self-employment income grew by an additional 4 percentage points.

Income imputation rates in the CPS grew more modestly than those in SIPP between 1984 and 1993 but then increased by 11 percentage points between 1993 and 2002. Imputation rates for all but two sources increased by about the same amount. The exceptions were property income, for which the imputation rate increased by 22 percentage points to 62.6 percent, and SSI, for which the increase was only 5 percentage points.

Quality of Imputation

The growing share of income that is imputed in these surveys makes it increasingly important that the imputations be done well. Both SIPP and the CPS have relied heavily on flexible hot-deck imputation procedures to impute missing items. Hot-deck imputation procedures replace missing values with values selected from other records—called donors—that are matched on a prespecified set of characteristics that form a large table. Flexible hot-deck procedures can combine the cells of a table, as necessary, to find donors when many of the cells are empty. Nevertheless, when item nonresponse is high—as it is for income and assets—the amount of collapsing that may be required to achieve matches reduces the quality of the imputations.

While the hot-deck algorithms that the Census Bureau employs can incorporate a large number of potentially relevant variables, the variables used to match donors to the records being imputed are not tailored, generally, to the items being imputed. For example, Doyle and Dalrymple (1987) demonstrated that by not taking into account reported Food Stamp Program benefits when imputing major components of income or by not taking account of income eligibility limits when imputing FSP benefits, the Census Bureau was imputing FSP benefits to households with incomes well beyond the eligibility limits or imputing high incomes to households that reported the receipt of FSP benefits. In response, the Census Bureau made improvements to address this particular problem as well as other related problems.

With the 1996 redesign and the need to rewrite numerous programs to run on the expanded, reformatted file, some of these enhancements appear to have been lost. In January 2003, for example, SIPP estimated that more than 400,000 adult FSP participants were in families with incomes four times the poverty level. FSP receipt was imputed to 62 percent of these persons compared with less than 7 percent of the estimated 6.3 million FSP participants with family incomes below poverty (Beebout and Czajka, 2005). This suggests that the Census Bureau is not taking sufficient account of income when imputing FSP receipt. Similarly, \$1.1 billion in welfare income was imputed in SIPP to families in the top income quintile in 2002 (Czajka, Mabli, and Cody, 2008). More than a third of all imputed welfare dollars went to families in the top income quintile in that year. This is comparable to only $10 million in welfare income imputed to the top income quintile in the CPS in the same year, or less than 1 percent of total imputed welfare dollars. In the years immediately preceding the 1996 redesign, the amounts of welfare income imputed to families in the top quintile were similar between SIPP and the CPS.

WAVE 1 BIAS

Since the redesign, each new SIPP panel (1996, 2001, and 2004) has started with a monthly poverty rate that was at least 2 percentage points higher than the poverty rate in the final wave of the preceding panel (Czajka, Mabli, and Cody, 2008). Undoubtedly, a number of factors contribute to this result, but one that has emerged with the most recent panels involves a possible understatement of income in Wave 1. Both the 1996 and 2001 panels showed a percentage point decline in the poverty rate between the first and second waves. In the 1996 panel, poverty continued to decline in the presence of an expanding economy, but in the 2001 panel there was no further decline in the poverty rate after the second wave. In the 2004 panel the Wave 1 to Wave 2 reduction was nearly 2 percentage points. Seasonal swings in income provide an obvious explanation, but the 1996 panel started 2 months later in the year than the 2001 and 2004 panels.

Panel surveys may be subject to a "time-in-sample" bias. Through repeated interviews, respondents may become better respondents as they learn what is expected of them. They may also become bored or learn how to avoid lengthy segments of the interview. Prior to the 1996 redesign, the Census Bureau compared data from overlapping waves in successive panels in a search for evidence of a time-in-sample bias in the reporting of income and benefit receipt in the SIPP. The research yielded no evidence of time-in-sample bias in SIPP (Lepkowski et al., 1992). With the elimination of overlapping panels, it is not possible to replicate this research on more recent SIPP data.

While there may be no evidence of a time-in-sample bias in earlier SIPP panels, there is a strong suggestion of some type of change in the reporting or perhaps processing of income data between the first two waves of more recent panels. Czajka, Mabli, and Cody (2008) compared poverty status between the first two waves of the 2004 panel in an effort to determine what role attrition and other sample loss might have played in the 1.8 percentage point decline in poverty. They found that changes in recorded poverty among persons present in both waves accounted for 87 percent of the net reduction in the number of poor between the two waves. Between Waves 2 and 3, a much smaller reduction in the number of poor (0.3 percentage points) could be attributed in large part to fewer gross exits from poverty—that is, fewer sample families reporting increased incomes.

Appendix B

Biographical Sketches of Panel Members and Staff

John Karl Scholz (*Chair*) is a professor of economics at the University of Wisconsin–Madison. Previously, he was the deputy assistant secretary for tax analysis at the U.S. Department of the Treasury and senior staff economist at the Council of Economic Advisers. He chaired the Committee on National Statistics' Panel on Enhancing the Data Infrastructure in Support of Food and Nutrition Programs, Research, and Decision-Making and was a member of its Panel on Evaluation of USDA's Methodology for Estimating Eligibility and Participation for the WIC Program. He has written extensively on the earned income tax credit and low-wage labor markets. He also writes on public policy and household saving, charitable contributions, and bankruptcy laws. He is a research associate at the National Bureau of Economic Research and was director of the Institute for Research on Poverty at the University of Wisconsin–Madison. He has a Ph.D. in economics from Stanford University.

F. Jay Breidt is professor and chair in the Department of Statistics at Colorado State University. He has published extensively in his research areas of time series, environmental monitoring, and survey sampling. From 1991-2000, he was on the faculty at Iowa State University. While at Iowa State, Breidt was a member of the Statistical Laboratory's Survey Section, which had as a major focus design and estimation for large-scale environmental surveys, particularly the USDA's National Resources Inventory. He is a fellow of the American Statistical Association and winner of the 2004 Distinguished Achievement Award from its Section on Statistics and the

Environment, as well as an elected member of the International Statistical Institute. He has M.S. and Ph.D. degrees from Colorado State University.

Leonard E. Burman is the Daniel Patrick Moynihan Chair in Public Affairs at the Maxwell School of Syracuse University. Prior to that, he directed the Tax Policy Center, which he founded with several colleagues in 2002. He was also a senior fellow at the Urban Institute and a visiting professor at Georgetown University. He served as deputy assistant secretary of the Treasury Department for tax analysis from 1998 to 2000, and as senior analyst at the Congressional Budget Office from 1988 to 1997. He is vice president of the National Tax Association, on the editorial board of *Public Finance Quarterly*, and a member of the International Monetary Fund Fiscal Analysis Division's Panel of Experts. He has served on several federal and local government advisory boards in the United States. He is the author of *The Labyrinth of Capital Gains Tax Policy: A Guide for the Perplexed*, and coeditor of *Taxing Capital Income and Using Taxes to Reform Health Insurance*. His recent research has examined the individual alternative minimum tax, the changing role of taxation in social policy, and tax incentives for savings, retirement, and health insurance. He has a Ph.D. from the University of Minnesota and a B.A. from Wesleyan University.

Constance F. Citro (*Study Director*) is the director of the Committee on National Statistics (CNSTAT), a position she has held since May 2004. She began her career with CNSTAT in 1984 as study director for the panel that produced *The Bicentennial Census: New Directions for Methodology in 1990*. Previously she held positions as vice president of Mathematica Policy Research, Inc., and Data Use and Access Laboratories, Inc. She was an American Statistical Association/National Science Foundation/Census research fellow and is a fellow of the American Statistical Association and an elected member of the International Statistical Institute. For CNSTAT, she directed evaluations of the 2000 census, the Survey of Income and Program Participation, microsimulation models for social welfare programs, and the National Science Foundation science and engineering personnel data system, in addition to studies on institutional review boards and social science research, estimates of poverty for small geographic areas, data and methods for retirement income modeling, and alternative poverty measures. She has a B.A. in political science from the University of Rochester and M.S and Ph.D. degrees in political science from Yale University.

John L. Czajka is a senior fellow at Mathematica Policy Research, Inc. His work has focused on the evaluation of estimates obtained from survey data and statistical uses of program administrative records. He has also directed many studies of health insurance coverage, including analyses of the dynamics of coverage over time and the impact of the Children's Health Insurance Program on trends in children's coverage. He has served on three previous National Academies panels, addressing issues related to the 2000 census, the adequacy of existing data for evaluating the impact of welfare reform, and setting priorities for research and development for the Census Bureau's state and local government statistics program. Prior to joining Mathematica in 1978, Dr. Czajka lectured at the University of California, Berkeley, in the Department of Sociology. Dr. Czajka is a past president of the Washington Statistical Society and a fellow of the American Statistical Association. He received his Ph.D. in sociology from the University of Michigan in 1979.

Peter Gottschalk is professor of economics in the School of Arts and Sciences at Boston College. His research interests cover labor economics and human resource economics. During the past decade, he has twice served as a visiting scholar at the Russell Sage Foundation, in fall 1993 and 1996. Before joining Boston College in 1987, he was associate professor of economics, professor of economics, and chairperson for 10 years at Bowdoin College. He has a B.A. in economics from the George Washington University and a Ph.D. in economics from the University of Pennsylvania.

Ronald T. Haskins is a senior fellow in the Economic Studies Program and codirector of the Center on Children and Families at the Brookings Institution and senior consultant at the Annie E. Casey Foundation in Baltimore. He is the author of *Work Over Welfare: The Inside Story of the 1996 Welfare Reform Law*, the coauthor of *Getting Ahead or Losing Ground: Economic Mobility in America*, and a senior editor of *The Future of Children*. In 2002 he was the senior adviser to the president for welfare policy at the White House. Prior to joining Brookings and Casey, he spent 14 years on the staff of the House Ways and Means Human Resources Subcommittee, first as welfare counsel to the Republican staff, then as the subcommittee's staff director. While there he edited the 1996, 1998, and 2000 editions of the *Green Book*. In 1981-1985, he was a senior researcher at the Frank Porter Graham Child Development Center at the University of North Carolina, Chapel Hill. He has a B.A. in history, an M.A. in education, and a Ph.D. in developmental psychology from the University of North Carolina, Chapel Hill.

V. Joseph Hotz is the arts and sciences professor of economics at Duke University. His research interests are in labor economics, economic demography, and evaluation of the impact of social programs. He has served on several National Academies panels, including the Panel on Transforming Our Common Destiny: Hispanics in the United States; the Panel on Access to

Research Data; the Panel on Institutional Review Boards, Surveys, and Social Science Research; the Panel to Evaluate the Survey of Income and Program Participation; and the Panel on Performance Measures for Data and Public Health Performance Partnership Grants. He is a member of the Committee on National Statistics. He has a Ph.D. in economics from the University of Wisconsin–Madison.

John Iceland is professor of sociology and demography at Pennsylvania State University. Previously, he was an associate professor of sociology at the University of Maryland–College Park and a faculty associate of the Maryland Population Research Center. He was chief of the Poverty and Health Statistics Branch at the U.S. Census Bureau before joining the Maryland faculty in 2003. His research focuses on poverty and residential segregation issues. His book, *Poverty in America*, is now in its second edition. He has authored numerous papers and reports on poverty patterns, causes, and measurement. His work on residential segregation examines general trends among various groups using a variety of measures, and he is currently examining the residential patterns of immigrants. He has a Ph.D. from Brown University.

Caryn Kuebler is on the staff of the U.S. Government Accountability Office. Prior to that, she served as an associate program officer with the Committee on National Statistics at the National Academies. She previously worked for the University of Chicago's Cultural Policy Center on a research project measuring the relationship between the size and scope of a region's creative sector and its economic growth potential. Her research interests include measuring consumer debt burden and income inequality, economic development, and cultural policy, including access to and protection of cultural and natural resources. She received her B.S. from Syracuse University and her M.P.P. from the University of Chicago.

Jerome P. Reiter is associate professor of statistical science at Duke University. His primary research focus has been investigating statistical methods of preserving the confidentiality of data. He works extensively on developing the theory and assessing the feasibility of releasing synthetic, that is, simulated, data to the public. He also develops methods for handling missing data in surveys and for inferring causal effects in observational studies. He has analyzed data from business, education, medicine, political science, psychology, public health, and sports. He has a Ph.D. in statistics from Harvard University.

Patricia Ruggles is a statistician in the Environmental-Economic Accounts Section of the United Nations Statistics Division. Prior to that, she served as a researcher and study director with the Committee on National Statistics (CNSTAT) and the Center on Economics, Governance and International Studies at the National Academies. She previously served as the Democratic staff director of the Joint Economic Committee, U.S. Congress; the chief economist and policy director for the Democratic staff of the Budget Committee of the U.S. House of Representatives; and as the deputy assistant secretary for income security policy and the chief economist at the Department of Health and Human Services, among other positions. Her research has focused on income, poverty, and anti-poverty policies, as well as on data and measurement issues. She served as a member of the CNSTAT Panel to Evaluate the Survey of Income and Program Participation in the early 1990s. Dr. Ruggles received her B.A. in history from Yale University and her Ph.D. in economics from Harvard.

Deanna T. Schexnayder is associate director and research scientist at the Ray Marshall Center for the Study of Human Resources, an organized research unit of the LBJ School of Public Affairs at the University of Texas at Austin. She has directed or codirected research projects on Texas welfare, education and training, and child care and child support programs for over two decades. She currently directs several large-scale research projects, including (with Christopher King) the Central Texas Student Futures Project, a regional initiative designed to study education and labor force outcomes for graduates in 10 central Texas school districts. She recently completed a multiyear assessment of the Texas subsidized child care system and is currently a member of a national research team that is designing a national survey of child care supply and demand for the assistant secretary for planning and evaluation of the U.S. Department of Health and Human Services. She is the coauthor (with Laura Lein) of *Life After Welfare: Reform and the Persistence of Poverty*. Since 1985, she has supervised the assembly of complex statistical research data sets from confidential, individual-level administrative data files from 16 different programs in Texas and other states. She has a B.S. in psychology and an M.B.A. from Louisiana State University.

Robert F. Schoeni is professor of economics and public policy and research professor at the Institute for Social Research, University of Michigan. His teaching and research interests include program evaluation, welfare policy, the economics and demographics of aging, labor economics, and immigration. He also serves as associate director of the Panel Study of Income Dynamics. He worked previously at RAND, where he was associate director of the labor and population program, and also served as senior economist on the President's Council of Economic Advisers. He has a Ph.D. in economics from the University of Michigan.

Jennifer Van Hook is an associate professor of sociology and demography at the Pennsylvania State University. She was previously associate professor of sociology at Bowling Green State University. Her research focuses on the health and well-being of the children of immigrants, including patterns and trends in household and family structure, child poverty, welfare receipt, food security, and child obesity and on the relationship between the policy contexts of reception and the incorporation patterns of immigrants and their children. She is also engaged in applied research that seeks to evaluate and revise methods for estimating the size, growth, and characteristics of the unauthorized migrant population living in the United States. She has an M.S. in sociology from the University of Wisconsin and a Ph.D. in sociology from the University of Texas.

References

Abowd, John (2007). *Assessing Disclosure Risk and Analytical Validity for the SIPP-SSA-IRS Public Use File Beta Version 4.1.* PowerPoint presentation to the Panel on the Census Bureau's Reengineered SIPP, Committee on National Statistics. Cornell University. Available: http://www.census.gov/sipp/dews.html [accessed May 15, 2009].

_____ (2008). *Assessing the Utility of Statistical Methods for Limiting Disclosure Risk: Value of Synthetic Data Sets.* PowerPoint presentation to the Workshop on Collecting, Storing, Protecting, and Accessing Biological Data Collected in Social Surveys, National Research Council. Cornell University (November 18).

An, D., and Roderick J.A. Little (2007). Multiple imputation: An alternative to top coding for statistical disclosure control. *Journal of the Royal Statistical Society, Series A* 170:923-940.

Assistant Secretary for Planning and Evaluation (2005). *Receipt of Unemployment Insurance Among Low-Income Single Mothers.* ASPE Issue Brief. Washington, DC: U.S. Department of Health and Human Services. Available: http://aspe.hhs.gov/hsp/05/unemp-receipt/index.htm [accessed May 15, 2009].

_____ (2008). *Indicators of Welfare Dependency: Annual Report to Congress 2008.* Washington, DC: U.S. Department of Health and Human Services. Available: http://aspe.hhs.gov/hsp/indicators08/index.shtml [accessed May 16, 2009].

Aughinbaugh, Allison, and Rosella M. Gardecki (2007). Attrition in the National Longitudinal Survey of Youth 1997. In *Proceedings of the FCSM Research Conference.* Washington, DC: Federal Committee on Statistical Methodology.

Beebout, Harold, and John L. Czajka (2005). Improving our ability to estimate the impact of changes to the transfer system and measure economic well-being: The legacy of Pat Doyle's early work. In *Proceedings of the Joint Statistical Meetings.* Alexandria, VA: American Statistical Association.

Belli, Robert F. (1998). The structure of autobiographical memory and the event history calendar: Potential improvements in the quality of retrospective reports in surveys. *Memory* 6:383-406.

_____ (2007). *Examining the Impact of Event History Calendar Interviewing on Data Quality from Disadvantaged Respondents.* Prepared for Census-PSID Event History Conference, Washington, DC, and Ann Arbor: Institute for Social Research, University of Michigan (December). Available: http://psidonline.isr.umich.edu/Publications/Workshops/ehc-07papers/Belli_Census_EHC_Conference_final.pdf [accessed March 28, 2009].

Brown, Clair, John Haltiwanger, and Julia Lane (2006). *Economic Turbulence—Is a Volatile Economy Good for America?* Chicago: University of Chicago Press.

Butricia, Barbara A., Howard M. Iams, and Karen E. Smith (2003). *It's All Relative: Understanding the Retirement Prospects of Baby Boomers.* Washington, DC: Urban Institute.

Callegaro, Mario (2008). Seam effects in longitudinal surveys. *Journal of Official Statistics* 24(3, September):387-429.

Callegaro, Mario, and Robert F. Belli (2007). *Impact of the Event History Calendar on Seam Effects in the PSID: Lessons for SIPP.* Prepared for Census-PSID Event History Conference, Washington, DC, and Ann Arbor: Institute for Social Research, University of Michigan (December). Available: http://psidonline.isr.umich.edu/Publications/Workshops/ehc-07papers/Callegaro_Belli_EHC_Census_conference_Submission.pdf [accessed March 28, 2009].

Carini, Robert M., John C. Hayek, George D. Kuh, John M. Kennedy, and Judith M. Ouimet (2003). College student responses to web and paper surveys. *Research in Higher Education* 44(1, February):1-19. Available: http://nsse.iub.edu/pdf/mode.pdf [accessed March 28, 2009].

Citro, Constance F. (2007). *Opportunities and Challenges for the Reengineered SIPP.* PowerPoint presentation to the Joint Statistical Meetings, Salt Lake City, UT. Committee on National Statistics, National Research Council, Washington, DC (August).

Coder, John F. (1988). Comparisons of Alternative Annual Estimates of Wage and Salary Income from SIPP. Memorandum for Gordon Green, assistant division chief, Population Division, U.S. Census Bureau, Washington, DC (March 29).

_____ (1992). Using administrative record information to evaluate the quality of income data collected in the Survey of Income and Program Participation. In *Proceedings of Statistics Canada Symposium 92—Design and Analysis of Longitudinal Surveys.* Ottawa: Statistics Canada.

Coder, John F., and Lydia Scoon-Rogers (1996). *Evaluating the Quality of Income Data Collected in the Annual Supplement to the March Current Population Survey and the Survey of Income and Program Participation.* SIPP Working Paper 215, U.S. Census Bureau. Washington, DC: U.S. Department of Commerce (July). Available: http://www.sipp.census.gov/sipp/workpapr/wp215.pdf [accessed June 5, 2009].

Congressional Budget Office (2003). *How Many People Lack Health Insurance and for How Long?* (May 12). Washington, DC. Available: http://www.cbo.gov/ftpdocs/42xx/doc4211/05-12-03-UninsuredBrief.pdf [accessed June 29, 2009].

Czajka, John L. (2007). Re-engineering SIPP: Some thoughts from a long time user. In *Proceedings of the Joint Statistical Meetings.* Alexandria, VA: American Statistical Association.

Czajka, John L., and Gabrielle Denmead (2008). *Income Data for Policy Analysis: A Comparative Assessment of Eight Surveys.* Prepared for the Office of the Assistant Secretary of Planning and Evaluation, U.S. Department of Health and Human Services (December). Washington, DC: Mathematica Policy Research.

Czajka, John L., and James Mabli (2009). *Analysis of Transition Events in Health Insurance Coverage* (May). Washington, DC: Mathematica Policy Research.

Czajka, John L., and Julie Sykes (2006). *Health Insurance Coverage Dynamics Among Children and Nonelderly Adults, 1996-1999* (October). Washington, DC: Mathematica Policy Research.

Czajka, John L., Jonathan E. Jacobson, and Scott Cody (2003). *Survey Estimates of Wealth: A Comparative Analysis and Review of the Survey of Income and Program Participation* (August). Washington, DC: Mathematica Policy Research.

Czajka, John L., James Mabli, and Scott Cody (2008). *Sample Loss and Survey Bias in Estimates of Social Security Beneficiaries: A Tale of Two Surveys* (February). Washington, DC: Mathematica Policy Research.

Dahl, Molly W., and John Karl Scholz (2005). The National School Lunch Program and School Breakfast Program: New Evidence on Participation and Noncompliance. Unpublished paper, University of Wisconsin–Madison.

Davern, Michael, Gestur Davidson, Jeanette Ziegenfuss, Stephanie Jarosek, Brian Lee, Tzy-Chyi Yu, Timothy J. Beebe, Kathleen T. Call, and Lynn A. Blewett (2007). *A Comparison of the Health Insurance Coverage Estimates from Four National Surveys and Six State Surveys: A Discussion of Measurement issues and Policy Implications.* Final report to the U.S. Department of Health and Human Services, Assistant Secretary for Planning and Evaluation and Agency for Healthcare Research and Quality (November). Minneapolis: State Health Access Data Assistance Center, University of Minnesota School of Public Health.

Decesaro, Anne, and Jeffrey Hemmeter (2008). *Characteristics of Noninstitutionalized DI and SSI Program Participants.* Research and Statistics Note No. 2008-02, Office of Policy. Washington, DC: U.S. Social Security Administration. Available: http://www.ssa.gov/policy/docs/rsnotes/rsn2008-02.html [accessed June 29, 2009].

Decressin, Anja, Tomeka Hill, and Julia Lane (2006). *Employer Provided Health Insurance: What Can Be Learned from the Form 5500?* Final Report (January), U.S. Census Bureau. Washington, DC: U.S. Department of Commerce.

Doyle, Pat, and Robert Dalrymple (1987). The impact of imputation procedures on distributional characteristics of the low income population. In *Proceedings of the Census Bureau Third Annual Research Conference.* Washington, DC: U.S. Department of Commerce.

Doyle, Pat, and Sharon K. Long (1988). The impact of the unit of analysis on measures of serial multiple program participation. In *Individuals and Families in Transition: Understanding Change Through Longitudinal Data.* Paper presented at the Social Science Research Council Conference in Annapolis, Maryland, March 16-18. Washington, DC: U.S. Department of Commerce.

Duncan, George T., and Diane Lambert (1989). The risk of disclosure for microdata. *Journal of Business and Economic Statistics* 7:207-217.

Duncan, George T., Sally A. Keller-McNulty, and S. Lynne Stokes (2001). *Disclosure Risk vs. Data Utility: The R-U Confidentiality Map.* Technical report. Research Triangle Park, NC: National Institute of Statistical Sciences.

Eargle, Judith (1990). *Household Wealth and Asset Ownership: 1988.* Current Population Reports, Household Economic Studies, Series P-70, No. 22, U.S. Census Bureau. Washington, DC: U.S. Department of Commerce.

______ (2004). Sample Loss Rates through Wave 9 for the 2001 Panel. Memorandum for the SIPP Record 2004-01(March 2). U.S. Census Bureau, Washington, DC.

Federal Committee on Statistical Methodology (1994). *Report on Statistical Disclosure Limitation Methodology.* Statistical Policy Working Paper 22 [revised 2005]. (NTIS PB94-165305). Washington, DC: U.S. Office of Management and Budget.

Fellegi, Ivan P., and A.B. Sunter (1969). A theory for record linkage. *Journal of the American Statistical Association* 64:1183-1210.

Fields, Jason M., and Jeffrey C. Moore (2007). *Description of Plans for a SIPP Calendar Validation Study: Study Design and Analysis.* Prepared for Census-PSID Event History Conference, Washington, DC. Washington, DC: U.S. Census Bureau (December). Available: http://psidonline.isr.umich.edu/Publications/Workshops/ehc-07papers/Fields%20and%20Moore%20EHC-11-26-2007.pdf [accessed July 10, 2009].

Fisher, T. Lynn (2007). The impact of survey choice on measuring the relative importance of Social Security benefits to the elderly. *Social Security Bulletin* 67(2):55-64.

Fitzgerald, John, Peter Gottschalk, and Robert Moffitt (1998). Analysis of sample attrition in panel data: The Michigan Panel Study of Income Dynamics. *Journal of Human Resources* 33(2):251-299.

Food and Nutrition Service (1999). *Current Population Survey of Analysis of NSLP Participation and Income*. Nutrition Assistance Program Report Series, Office of Analysis, Nutrition, and Evaluation. Washington, DC: U.S. Department of Agriculture. Available: http://www.fns.usda.gov/ora/MENU/Published/CNP/FILES/nslpcps.pdf [accessed May 15, 2009].

Fuller, Wayne A. (1993). Masking procedures for microdata disclosure limitation. *Journal of Official Statistics* 9:383-406.

Gbur, Philip M., Patrick J. Cantwell, and Rita J. Petroni (1990). Effect of telephone interviewing on SIPP topical module and longitudinal estimates. Pp. 570-575 in *Proceedings of the Survey Research Methods Section*. Alexandria, VA: American Statistical Association.

Glenn, David (2006). Social scientists protest plan to end federal study of income and hardship. *Chronicle of Higher Education* March 2.

Goerge, Robert (2009). *Employment Outcomes for Low-Income Families Receiving Child Care Subsidies in Illinois, Maryland, and Texas*. Report to the U.S. Department of Health and Human Services. Chicago: University of Chicago.

Gomatam, S., Alan F. Karr, Jerome P. Reiter, and A.P. Sanil (2005). Data dissemination and disclosure limitation in a world without microdata: A risk-utility framework for remote access servers. *Statistical Science* 20:163-177.

Gottschalk, Peter, and Minh Huynh (2006). *Are Earnings Inequality and Mobility Overstated? The Impact of Non-Classical Measurement Error.* IZA Discussion Paper 2327. Available: http://papers.ssrn.com/sol3/papers.cfm?abstract_id=937340 [accessed June 5, 2009].

Hotz, V. Joseph, and John Karl Scholz (2002). Measuring employment and income outcomes for low-income populations with administrative and survey data. Pp. 275-315 in National Research Council, *Studies of Welfare Populations: Data Collection and Research Issues*. Panel on Data and Methods for Measuring the Effects of Changes in Welfare Programs, Michele Ver Ploeg, Robert A. Moffitt, and Constance Citro, Eds. Committee on National Statistics, Division of Behavioral and Social Sciences and Education. Washington, DC: National Academy Press.

Hotz, V. Joseph, Charles H. Mullin, and John Karl Scholz (2003). *Examining the Effect of the Earned Income Tax Credit on the Labor Market Participation of Families on Welfare*. Department of Economics, Duke University. Available: http://www.econ.duke.edu/~vjh3/working_papers/EITC2.pdf [accessed May 15, 2009].

________ (2005). *Trends in EITC Take-Up and Receipt for California's Welfare Population, 1992-1999*. Department of Economics, Duke University. Available: http://www.econ.duke.edu/~vjh3/working_papers/EITC3.pdf [accessed May 15, 2009].

Huggins, Vickie, and Robert Fay (1988). Use of administrative data in SIPP longitudinal estimation. Pp. 354-359 in *Proceedings of the Section on Survey Research Methods*. Alexandria, VA: American Statistical Association.

Huynh, Minh, Kalman Rupp, and James Sears (2001). The assessment of Survey of Income and Program Participation (SIPP) benefit data using longitudinal administrative records. In *Proceedings of the FCSM Research Conference*. Washington, DC: Federal Committee on Statistical Methodology. Also available as SIPP Working Paper No. 238 (2002): http://www.census.gov/sipp/workpapr/wp238.pdf [accessed May 15, 2009].

Iams, Howard M., and Steven H. Sandell (1996). Past is prologue: Simulating lifetime Social Security earnings for the twenty-first century. In *Proceedings of the U.S. Bureau of the Census 1996 Annual Research Conference* (March). Washington, DC: U.S. Department of Commerce.

Jabine, Thomas B., Karen E. King, and Rita J. Petroni (1990). *Survey of Income and Program Participation: Quality Profile.* U.S. Census Bureau. Washington, DC: U.S. Department of Commerce.

James, Tracey L. (1997). Results of the wave incentive experiment in the 1996 Survey of Income and Program Participation. In *Proceedings of the Joint Statistical Meetings* (August, Anaheim, CA). Alexandria, VA: American Statistical Association.

Johnson, David (2008). *The SIPP: The Evolution of a Phoenix.* PowerPoint presentation to the Population Association of America, New Orleans. U.S. Census Bureau, Washington, DC (April). Available: http://www.census.gov/sipp/dews.html [accessed May 15, 2009].

Kasprzyk, Daniel (1983). Social Security number reporting, the use of administrative records, and the multiple frame design in the Income Survey Development Program. Pp. 123-141 in Martin H. David, Ed., *Technical, Conceptual, and Administrative Lessons of the Income Survey Development Program (ISDP): Papers Presented at a Conference.* New York: Social Science Research Council.

______ (1988). Outline of questions in Committee on National Statistics proposal to evaluate the Survey of Income and Program Participation. Unpublished memorandum to Margaret E. Martin, U.S. Census Bureau, Washington, DC.

Kennickell, Arthur B. (2006). A rolling tide: Changes in the distribution of wealth in the U.S., 1989-2001. In Edward N. Wolff, Ed., *International Perspectives on Household Wealth.* Cheltenham, UK: Edward Elgar.

Koenig, Melissa L. (2003). An assessment of the Current Population Survey and the Survey of Income and Program Participation using Social Security administrative data. In *Proceedings of the FCSM Research Conference.* Washington, DC: Federal Committee on Statistical Methodology.

Lamas, Enrique, Jan Tin, and Judith Eargle (1994). *The Effect of Attrition on Income and Poverty Estimates from the Survey of Income and Program Participation.* Paper presented at the Conference on Attrition in Longitudinal Surveys. U.S. Census Bureau, U.S. Department of Commerce, Washington, DC.

Lambert, Diane (1993). Measures of disclosure risk and harm. *Journal of Official Statistics* 9:313-331.

Lepkowski, James M., Steven G. Pennell, David P. Miller, Elma Luis, and Graham Kalton (1992). *Time in Panel Effects in the SIPP.* Prepared for the U.S. Census Bureau, SIPP Working Paper No. 172. Ann Arbor, MI: Institute for Social Research, University of Michigan. Available: http://www.census.gov/sipp/workpapr/wp172.pdf [accessed September 4, 2009]

Little, Roderick J.A. (1993). Statistical analysis of masked data. *Journal of Official Statistics* 9:407-426.

Logan, Weltha J., Daniel Kasprzyk, and Roger Cavanaugh (1988). *A Methodological Study Using Administrative Records: The Special Frames Study of the Income Survey Development Program.* SIPP Working Paper No. 8814, U.S. Census Bureau. Washington, DC: U.S. Department of Commerce.

Lynch, Victoria, Dean M. Resnick, Jane Staveley, and Cynthia M. Taeuber (2008). *Differences in Estimates of Public Assistance Recipiency between Surveys and Administrative Records.* Prepared for the Family Investment Administration, Maryland Department of Human Resources (January). Washington, DC, and Baltimore, MD: U.S. Census Bureau and The Jacob France Institute, University of Baltimore. Available: http://www.jacob-france-institute.org/publications.php [accessed September 4, 2009].

Marquis, Kent, and Jeffrey Moore (1989). Response errors in SIPP: Preliminary results. Pp. 515-536 in *Proceedings of the Bureau of the Census Fifth Annual Research Conference.* Washington, DC: U.S. Department of Commerce.

_______ (1990a). Measurement errors in SIPP program reports. Pp. 721-745 in *Proceedings of the Bureau of the Census Sixth Annual Research Conference*. Washington, DC: U.S. Department of Commerce.

_______ (1990b). SIPP Record Check Final Report. Draft. Bureau of the Census, Washington, DC.

McKee, Nat, and Zelda McBride (2008). *SIPP Imputation Scheme and Discussion Items.* PowerPoint presentation to the ASA/SRM SIPP Working Group Meeting (September 16). U.S. Census Bureau, Washington, DC.

Meyer, Bruce, D. Wallace, K.C. Mok, and James X. Sullivan (2009). *The Under-Reporting of Transfers in Household Surveys: Its Nature and Consequences.* Working Paper 0903 (February). Chicago: Harris School of Public Policy Studies, University of Chicago.

Moore, J., J. Fields, J. Pascale, G. Benedetto, M. Stinson, and A. Chan (2009). *A Comparison of Survey Reports Obtained via Standard Questionnaire and Event History Calendar.* Paper presented at the Annual Meeting of the American Association for Public Opinion Research, Hollywood, FL, May 14-17.

National Research Council (1989). *The Survey of Income and Program Participation: An Interim Assessment.* Committee on National Statistics. Washington, DC: National Academy Press.

______ (1993). *The Future of the Survey of Income and Program Participation.* Panel to Evaluate the Survey of Income and Program Participation, Constance F. Citro and Graham Kalton, Eds. Committee on National Statistics, Commission on Behavioral and Social Sciences and Education. Washington, DC: National Academy Press.

______ (2001). *Evaluating Welfare Reform in an Era of Transition.* Panel on Data and Methods for Measuring the Effects of Changes in Social Welfare Programs, Robert A. Moffitt and Michele Ver Ploeg, Eds. Committee on National Statistics, Division of Behavioral and Social Sciences and Education. Washington, DC: National Academy Press.

______ (2004a). *Reengineering the 2010 Census: Risks and Challenges.* Panel on Research on Future Census Methods, Daniel L. Cork, Michael L. Cohen, and Benjamin F. King, Eds. Committee on National Statistics, Division of Behavioral and Social Sciences and Education. Washington, DC: The National Academies Press.

______ (2004b). *The 2000 Census: Counting Under Adversity.* Panel to Review the 2000 Census, Constance F. Citro, Daniel L. Cork, and Janet L. Norwood, Eds. Committee on National Statistics, Division of Behavioral and Social Sciences and Education. Washington, DC: The National Academies Press.

______ (2005). *Expanding Access to Research Data: Reconciling Risks and Opportunities.* Panel on Data Access for Research Purposes, Committee on National Statistics, Division of Behavioral and Social Sciences and Education. Washington, DC: The National Academies Press.

______ (2007). *Putting People on the Map: Protecting Confidentiality with Linked Social-Spatial Data.* Panel on Confidentiality Issues Arising from the Integration of Remotely Sensed and Self-Identifying Data, Committee on the Human Dimensions of Global Change. Division of Behavioral and Social Sciences and Education. Washington DC: The National Academies Press.

Newcombe, Howard B., J.M. Kennedy, S.J. Axford, and A.P. James (1959). Automatic linkage of vital records. *Science* 130:954-959.

Parker, Jennifer D., and Nathaniel Schenker (2007). Multiple imputation for national public-use datasets and its possible application for gestational age in U.S. natality files. *Paediatric and Perinatal Epidemiology* 21(s2):97-105.

Passel, Jeffrey S. (2006). *The Size and Characteristics of the Unauthorized Migrant Population in the United States.* Washington, DC: Pew Hispanic Center (March 7). Available: http://pewhispanic.org/reports/report.php?ReportID=61 [accessed May 15, 2009].

Pierret, Charles, Alison Aughinbaugh, A. Rupa Datta, and Tricia Gladen (2007). *Event History Data: Lessons from the NLSY97*. Prepared for Census-PSID Event History Conference, December, Washington, DC. Available: http://psidonline.isr.umich.edu/Publications/Workshops/ehc-07papers/Aughinbaugh%20EHC.pdf [accessed March 28, 2009].

Raghunathan, T.E., J.M. Lepkowski, J. van Hoewyk, and P. Solenberger (2001). A multivariate technique for multiply imputing missing values using a series of regression models. *Survey Methodology* 27:85-96.

Reese, Kanin L. (2007). *An Analysis of the Characteristics of Multiple Program Participation Using the Survey of Income and Program Participation (SIPP)*. SIPP Working Paper 246, U.S. Census Bureau. Washington, DC: U.S. Department of Commerce.

Reiter, Jerome P. (2003). Inference for partially synthetic, public use microdata sets. *Survey Methodology* 29:181-189.

_____ (2004). New approaches to data dissemination: A glimpse into the future. *Chance* 17(3):12-16.

_____ (2005). Estimating identification risks in microdata. *Journal of the American Statistical Association* 100:1103-1113.

Reiter, Jerome P., and R. Mitra (2009). Estimating risks of identification disclosure in partially synthetic data. *Journal of Privacy and Confidentiality* 1(1):99-110.

Reiter, Jerome P., A. Oganian, and Alan F. Karr (2009). Verification servers: Enabling analysts to assess the quality of inferences from public use data. *Computational Statistics and Data Analysis* 53:1475-1482.

Roemer, Marc I. (2000). *Assessing the Quality of the March Current Population Survey and the Survey of Income and Program Participation Income Estimates, 1990-1996*. Housing and Household Economic Statistics Division Working Paper. U.S. Census Bureau (July). Washington, DC: U.S. Department of Commerce. Available: http://www.census.gov/hhes/www/income/assess1.pdf [accessed June 5, 2009].

_____ (2002). *Using Administrative Earnings Records to Assess Wage Data Quality in the March Current Population Survey and the Survey of Income and Program Participation*. Longitudinal Employer-Household Dynamics Technical Paper No. TP-2002-22 (November), U.S. Census Bureau. Washington, DC: U.S. Department of Commerce.

Rubin, Donald B. (1987). *Multiple Imputation for Nonresponse in Surveys*. New York: John Wiley & Sons.

Sastry, Narayan, Anne Pebley, and Christine Peterson (2007). *The Design and Use of an Event History Calendar in the Los Angeles Family and Neighborhood Survey: Results and Lessons)*. Prepared for Census-PSID Event History Conference, December, Washington, DC. Available: http://psidonline.isr.umich.edu/Publications/Workshops/ehc-07papers/LAFANS%20EHC%20Paper%20Draft%202.pdf [accessed March 28, 2009].

Schafer, Joseph L. (1997). *Analysis of Incomplete Multivariate Data*. London: Chapman & Hall.

Schenker, Nathaniel, Trivellore E. Raghunathan, Pei-Lu Chiu, Diane M. Makuc, Guangyu Zhang, and Alan J. Cohen (2006). Multiple imputation of missing income data in the National Health Interview Survey. *Journal of the American Statistical Association* 101:924-933.

Scholz, John Karl (1994). The Earned Income Tax Credit: Participation, compliance, and anti-poverty effectiveness. *National Tax Journal* March:59-81.

Stinson, Martha (2008). *Research on Improvements to Current SIPP Imputation Methods*. PowerPoint presentation to the ASA-SRM SIPP Working Group (September), U.S. Census Bureau, Washington, DC. Available: http://www.census.gov/sipp/dews.html [accessed May 15, 2009].

Sylvester, Douglas J., with Kirk J. Bardin and Jessica M. Wann (2008). *Final Report and Findings: Legal Analysis of Proposed "DEWS" Program and Access to State-Held Records.* Prepared for the Committee on National Statistics Panel on the Census Bureau's Reengineered SIPP (March 2). Center for the Study of Law, Science, and Technology, Arizona State University, Tempe.

Taeuber, Cynthia, Dean M. Resnick, Susan P. Love, Jane Stavely, Parke Wilde, and Richard Larson (2004). *Differences in Estimates of Food Stamp Program Participation Between Surveys and Administrative Records.* Working Paper. Washington, DC: U.S. Census Bureau.

University of Texas and U.S. Census Bureau (2008). Agreement Between Ray Marshall Center for the Study of Human Resources at the University of Texas at Austin (RMC) and the U.S. Census Bureau for Use of Confidential Texas Administrative Records (June). Unpublished.

Urban Institute (2006). *Children of Immigrants: Facts and Figures. Fact Sheet,* The Urban Institute, Washington, DC (May). Available: http://www.urbaninstitute.org/UploadedPDF/900955_children_of_immigrants.pdf [accessed May 16, 2009].

Urban Institute/NORC Evaluation Team (2009). *SSA/SIPP/IRS Synthetic Beta File Analytic Evaluation.* Working Papers. Prepared for the Social Security Administration (March 31). Washington, DC: The Urban Institute.

U.S. Census Bureau (1998). *SIPP Quality Profile*, 3rd edition. SIPP Working Paper No. 230. Prepared by Graham Kalton, Westat, Inc., for the U.S. Census Bureau. Washington, DC: U.S. Department of Commerce.

______ (2002). *Current Population Survey Technical Paper 63RV: Design and Methodology* (March). Washington, DC: U.S. Department of Commerce.

U.S. General Accounting Office (1999). *Survey Methodology—An Innovative Technique for Estimating Sensitive Survey Items.* Staff Study. Washington, DC: U.S. Government Printing Office (November). Available: http://www.gao.gov/archive/2000/gg00030.pdf [accessed May 17, 2009].

Vaughan, Denton R. (1993). Reflections on the income estimates from the initial panel of the Survey of Income and Program Participation (SIPP). *Studies in Income Distribution*, No. 17 (May). Washington, DC: Social Security Administration.

______ (2007). *Factors that Facilitated and Inhibited Job-holding Among Female AFDC/TANF Recipients in 1996.* SIPP Working Paper No. 247. Prepared for the Housing and Household Economic Statistics Division, U.S. Bureau of the Census. Washington, DC: U.S. Department of Commerce. Available: http://www.census.gov/sipp/workpapr/wp247.pdf [accessed June 5, 2009].

Vaughan, Denton, and Fritz Scheuren (2002). Longitudinal attrition in SIPP and SPD. In *Proceedings of the Social Statistics Section.* Alexandria, VA: American Statistical Association.

Wagner, Deborah (2007). *Person SSN Validation System (PVS).* PowerPoint presentation to a working group of the Committee on National Statistics Panel on the Census Bureau's Reengineered SIPP (March 7). U.S. Bureau of the Census, Washington, DC.

Warner, Stanley L. (1965). Randomized response: A survey technique for eliminating evasive answer bias. *Journal of the American Statistical Association* 60:63-69.

Wheaton, Laura (2007). *Underreporting of Means-Tested Transfer Programs in the CPS and SIPP.* Paper presented to the Joint Statistical Meetings, Salt Lake City, UT (August). The Urban Institute, Washington, DC.

Winkler, William E. (2006). *Overview of Record Linkage and Current Research Activities.* Research Report Series #2006-2. Statistical Research Division, U.S. Census Bureau. Washington, DC: U.S. Department of Commerce.

Wolff, Edward N. (1999). The size distribution of wealth in the United States: A comparison among recent household surveys. In James P. Smith and Robert J. Willis, Eds., *Wealth, Work, and Health: Innovations in Measurement in the Social Sciences. Essays in Honor of F. Thomas Juster*. Ann Arbor: University of Michigan Press.

Ycas, Martynas A., and Charles A. Lininger (1983). The Income Survey Development Program: Design features and initial findings. Pp. 25-36 in Martin David, Ed., *Technical, Conceptual, and Administrative Lessons of the Income Survey Development Program (ISDP): Papers Presented at a Conference*. New York: Social Science Research Council.

Zabel, Jeffrey E. (1998). An analysis of attrition in the Panel Study of Income Dynamics and the Survey of Income and Program Participation with an application to a model of labor market behavior. *Journal of Human Resources* 33(2):479-506.

Zedlewski, Sheila, and Linda Giannarelli (2001). *Diversity Among State Welfare Programs: Implications for Reform. New Federalism: Issues and Options for States*. Washington, DC: Urban Institute.

Zimmermann, Wendy, and Laura Tumlin (1998). *Immigrants and Public Benefits: State Policies and Trends*. Washington, DC: Urban Institute.

COMMITTEE ON NATIONAL STATISTICS

The Committee on National Statistics (CNSTAT) was established in 1972 at the National Academies to improve the statistical methods and information on which public policy decisions are based. The committee carries out studies, workshops, and other activities to foster better measures and fuller understanding of the economy, the environment, public health, crime, education, immigration, poverty, welfare, and other public policy issues. It also evaluates ongoing statistical programs and tracks the statistical policy and coordinating activities of the federal government, serving a unique role at the intersection of statistics and public policy. The committee's work is supported by a consortium of federal agencies through a National Science Foundation grant.